Harris Hawthorne Wilder

A synopsis of the animal kingdom together with a laboratory practicum of invertebrate zoölogy

Harris Hawthorne Wilder

A synopsis of the animal kingdom together with a laboratory practicum of invertebrate zoölogy

ISBN/EAN: 9783337175047

Printed in Europe, USA, Canada, Australia, Japan

Cover: Foto ©berggeist007 / pixelio.de

More available books at **www.hansebooks.com**

A

SYNOPSIS OF THE ANIMAL KINGDOM,

TOGETHER WITH A

LABORATORY PRACTICUM

OF

INVERTEBRATE ZOÖLOGY.

BY HARRIS H. WILDER.

—

Privately printed for the use of students in Smith College.

—

Northampton, Mass.,
The Gazette Printing Company,
1894.

S

To my students in '95 and '96, in acknowl-
edgement of the patient work with which
they have aided me in the completion of
this book.

CONTENTS.

I. SYNOPSIS.

.

SYNOPSIS OF TYPES.

A. PROTOZOA.

B. METAZOA.

DIVISION I. PROTAXONIA (*Coelenterata*).

TYPE I. PORIFERA.

TYPE II. CNIDARIA.

TYPE III. CTENOPHORA.

DIVISION II. HETERAXONIA (*Coelomata*).

TYPE IV. ZYGONEURA.

SUB-TYPE I. PROTONEPHRIDOZOA (*Scolecida*).

Cladus I. Platodes.

Cladus II. Vermes.

SUB-TYPE II. METANEPHRIDOZOA (*Aposcolecida*).

Cladus I. Articulata.

Sub-cladus I. Annelida.

Sub-cladus II. Arthropoda.

Cladus II. Mollusca.

Cladus III. Molluscoidea.

TYPE V. AMBULACRALIA.

Cladus. Echinodermata.

TYPE VI. CHORDATA.

Cladus I. Hemichorda.

Cladus II. Urochorda (*Tunicata*).

Cladus III. Cephalochorda (*Leptocardii*).

Cladus IV. Vertebrata.

DEFINITIONS OF TYPES.

A. PROTOZOA.

Unicellular animal organisms. Reproduction by fission and gemmation, also by conjugation, which may simulate sex-differentiation. Colonial forms may show polymorphism (with division of labor) and thus serve as transition-forms between PROTOZOA and METAZOA.

B. METAZOA.

Multicellular animal organisms with sex-differentiation. The individual begins as a single cell, formed by the union of two half cells (egg and sper-matozoön) produced by the parents.

During development all metazoa pass through a *blastula* and *gastrula* stage, supposed repetitions of ancient *blastaea* and *gastraea* forms.

DIVISION I. PROTAXONIA (*Coelenterata*).

"Gastraea animals," i. e. forms with persisting gastraea characteristics, the primary axis, the protostoma and the gastrocoele, from which all the cavities of the body may be derived. No coelom.

TYPE I. PORIFERA.

Gastraea animals, sessile by the protostoma. Secondary excurrent opening, *osculum*, at the apical pole; with numerous lateral pores in the body wall. Middle layer (mesenchyma) present, derived from the primary endoderm. In this are produced spicules, and the germ cells.

TYPE II. CNIDARIA.

Gastraea-animals, mostly marine, free-swimming or sessile at the aboral pole. Develop tentacles around the protostoma. Certain specialized ecto-derm cells form Cnidoblasts (nettle cells). These are most numerous on the tentacles and serve as weapons. Middle layer a supporting lamella, seldom cellular. Germ cells from ectoderm or endoderm.

TYPE III. CTENOPHORA.

Gastraea animals, all marine : structure a modified radiate, or doubly bi-lateral one. Free-swimming : moving by means of eight meridional rows of peculiar organs derived from cilia. Sense organs at apical pole. Adhesive cells with spiral, contractile thread developed in ectoderm, replacing the Cni-doblasts of Type II. Middle layer a richly developed gelatinous tissue filled with muscle cells. Germ cells probably from ectoderm (at present a matter of controversy).

[In comparing the three groups of Gastraea-animals, Types I-III, notice the differences in development and use of the apical pole. In I it serves as an excurrent orifice ; in II as point of support, and in III it is directed forward during locomotion and develops sense-organs.]

DIVISION II. HETERAXONIA (*Coelomata*).

Bilateral animals, primary axis either embryonic or suppressed. Secondary (middle) layer from two sources : (1) from paired epithelial sacks (=*mesoderm*); (2) from wandering cells, derived from the primary endoderm (=*mesenchyma*). The first is a new formation and is often reduced to single cells, or otherwise obscurely indicated in the lower forms. The second is probably homologous with the middle layer of the Protaxonia. Coelom present, either a proto-coelom (derivative of Blastocoele) or metacoelom (cavities of the mesodermic sacks).

TYPE IV. ZYGONEURA.

Animals derived from the hypothetical *Trochozoön* (or an earlier form, *Pro-trochozoön*) which are frequently repeated throughout the class in the "Trochophore-larva" and similar forms. Nervous system consists of an api-cal cerebral ganglion from which proceeds at least one pair of longitudinal nerves.

SUB-TYPE I. PROTONEPHRIDOZOA (*Scolecida*).

Entire body a "prosoma" (i. e. derived from the ancestral form), main body cavity the protocoelom. Metacoelom represented as protonephridium and pri-mary sack-gonads.

Cladus I. Platodes.

Forms derived from the *Protrochozoön*, which is represented in a few mem-bers of the group by a "protrochula" larva. The adult forms are flat, either creeping or swimming slowly by means of cilia, or reduced by parasitic life.

Intestine incomplete and without anal orifice. Protocoelom secondarily filled up by a connective tissue parenchyma, derived from the mesenchyma. Protonephridium a system of branching tubules, the so-called "water-vascular system." Mostly hermaphroditic with complicated reproductive system.

Cladus II. Vermes.

Forms derived from *Trochozoön*. Protocoelom serves as body cavity and is not filled with mesenchyma. Intestine complete with anus. The animals of this class are externally very unlike and fitted for many different environments. Some are plainly modified Trochophores when adult, others pass this stage in early life, while in others the stage has dropped out. They are most easily distinguished from other worm-like forms by their lack of segmentation.

SUB-TYPE II. METANEPHRIDOZOA (*Aposcolecida*).

Anterior part of body alone (*Prosoma*) derived from Trochozoön, to which the remainder (=*Metasoma*, generally the larger portion) is added by successive growths forming segments or somites. Each somite forms organs like the prosoma, which are distinguished as *meta-nephridia*, and *secondary gonads*. In some forms the organs of the prosoma remain to supply that portion, in others they early disappear, the region being secondarily supplied by ingrowths from the metasomatic organs. The body-cavity is the meta-coelom, formed by paired mesodermic coelom sacks, which may be considered as expansions of the sack gonads in which the germ cells are localized in certain areas as "surface gonads." The metanephridia communicate internally with the coelom and externally with the outside. They may be repetitions or segmentations of the protonephridia, secondarily connected with the coelom : or they may be evaginations of the coelom itself.

Cladus I. Articulata.

Segmentation very pronounced, each segment typically furnished with a pair of appendages, jointed in the higher forms. Segments and appendages often show marked specialization in different body regions. Integument protected by a cuticula which may become a thick chitinous exoskeleton. Metacoelom spacious, mesenchyma poorly developed ; nervous system a double cerebral ganglion dorsal to intestine, connected by commissures to a ventral ganglionic chain. Metanephridia in the typical form segmentally arranged, in higher forms modified or fail, being replaced by other organs.

Sub-cladus I. Annelida.

Appendages in the form of *parapodia*, i. e. fleshy outgrowths of the lateral walls, bearing bristles, gills, and other organs, but never jointed. In some forms the processes are suppressed, the bristles being inserted in pits. No marked differentiation of the body into definite body regions. In this group appear the most typical Trochophore larvae.

Sub-cladus II. Arthropoda.

Appendages jointed. A Trochophore larva does not appear, but their position is determined by means of their marked relationship to the Annelids. Development by larval forms is common, but many of these are plainly secondarily acquired and of no value in phylogenesis.

Cladus II. Mollusca.

Metasoma consisting of one segment, which early fuses with the prosoma, resulting in an apparently unsegmented body. Metanephridia a single pair. Metacoelom almost filled with parenchymatous tissue, the pericardial cavity and those of the germ glands alone remaining. The body consists typically of head, foot and visceral sack, of which the first two may fail or be modified beyond recognition. The body is generally surrounded by a duplicature of the integument termed the mantle, which secretes a calcareous shell and encloses a branchial cavity. As reminiscence of the Trochozoön, there frequently appears the "Veliger" larva, similar to the Trochophore.

Cladus III. Molluscoidea.

Sessile animals, mostly marine and often colonial, either enclosed in cells or protected by dorsal and ventral calcareous shells. Mouth surrounded by a ridge, the *lophophore*, which bears ciliated tentacles, or two spiral ciliated arms. Intestine U-shaped, with anus just outside the lophophore. Metacoelom present, developing in the typical forms from a pair of lateral sacs. One pair of metanephridia. Central nervous system a ganglion between mouth and anus. The apparent lack of segmentation may be explained here as in Mollusca. by supposing a simple metasoma

INVERTEBRATE ZOÖLOGY. 13

TYPE V. AMBULACRALIA (with one Cladus).

Cladus. Echinodermata.

Primarily bilateral pelagic forms, which attain a secondary radiate penta-merous structure, after which they may become creeping or even sessile. The bilateral larvae are all derivable from a common form, the *Pro-auricularia*, which is itself derived directly from the gastraea by the formation of a sec-ondary mouth and the conversion of the protostoma into the anus. This form somewhat resembles the trochozoön, but differs from it in the origin of mouth and anus, and in the absence of an apical sense organ.

During development the protocoelom becomes filled with mesenchyma, and the definite coelom appears in the form of a pair of mesodermic sacs (or one double sac), the *enterocoeles*, from which arise typically a pair of secondary sacs, assuming the form of five-pointed rosettes, the *hydrocoeles*. Of these, the right one aborts and the left one grows into the water-vascular system of the definite radiate form. In most cases the right hydrocoele is entirely sup-pressed. The adult form possesses oral and ab-oral surfaces, which corres-pond respectively to the left and right sides of the posterior portion of the larva.

Larval locomotion is by cilia, the adult mainly by the water-vascular sys-tem. Characteristic in all stages is the possession of a skeleton, composed of calcareous elements, which may form closely articulated plates, a network of loosely woven spicules, or may be reduced to minute granules imbedded in the integument.

[P. and L. SARASIN and others attribute to the hydrocoeles the value of nephridia, and BURY claims that the enterocoele normally develops from *two* pairs of coelom-sacs.]

TYPE VI. CHORDATA.

Bilateral animals with dorsal nervous system, supported ventrally during some stage at least by a skeletal rod, the *notochord*, derived from the endo-derm. The pharynx is perforated by lateral slits, respiratory in function, which may develop gills, supported by skeletal rods, the visceral or gill arches. The protostoma is embryonic and occupies the future dorsal region, appearing in higher embryoes as an elongated groove or cleft, the *primitive streak*. The mesoderm arises from the lateral region of the primordial intestine, typically in the form of paired sacs, segmentally arranged, but in modified forms as more or less solid masses, which segment and develop cavities after separation from the endoderm. From these mesodermic cavities the metacoelom is formed. The protocoelom becomes filled with mesenchyma, which forms the connective tissues and the blood vessels.

Cladus I. Hemichorda.

Notochord confined to anterior part of body above pharynx. Nervous system dorsal and ventral chords, not entirely distinct from ectoderm. Pharynx perforated by paired gill slits. Genital organs consist of several pairs of sac-gonads, opening independently on the dorsal side. As diverticula of the primordial intestine (*archenteron*) arise 2-3 pairs of coelom-sacs, as well as a median dorsal pore, compared by some to the dorsal pore of Echinoderm larvae. This appears double in some species. Some species develop directly, others by a free-swimming larva, the *Tornaria*, which appears in some respects similar to Proauricularia, and in others to the trochophore. Development by Tornaria-larva, is undoubtedly more primitive than the direct method, and hence Tornaria must represent the free-swimming pelagic ancestor of the Chordata. Its relationship to the other primitive forms is uncertain.

Cladus II. Urochorda (*Tunicata*).

A group of marine Chordata, representing various stages of degeneracy, and thus of very varied external appearance. The larvae are free-swimming and pelagic, but may become sessile in adult life. The notochord is confined to the caudal region where it supports the posterior part of the dorsal nervous system. It is persistant throughout life in a single group of small pelagic forms, the *Perennichordata* (*Appendicularia*), becoming reduced in all others. Corresponding to the changes in the notochord, the larval nervous system is elongated, but in the adult becomes reduced to a single ganglion, lying dorsal to the pharynx. The mesoderm arises as in typical chordata, but is without trace of segmentation, and lacks the coelomic cavities, which either do not appear or become early obliterated. Nephridia fail. The body of the adult is enclosed in a mantle or tunic, which may be gelatinous, leathery or cartilaginous, and posesses two orifices : an incurrent or oral, and an excurrent or cloacal. The pharynx is perforated by gill slits which may appear in two lateral rows, or by secondary reduplication may convert the entire pharyngeal wall into a lattice-work. The water passes from these slits into a peribranchial or cloacal cavity, formed by the mantle, and into which the intestine and the genital ducts also open. In the sessile forms the intestine is U-shaped, and the two mantle openings approximate, while in the free-swimming forms they are situated at the two poles of the oval or cask-shaped body. Asexual reproduction combined with complicated alternation of generations occur in this group. The sexual forms are hermaphroditic.

Cladus III. Cephalochorda.

Elongated fish-like marine forms, without head, brain, or skull. Notochord persistant, extending through entire body from tip to tip. Mouth near anterior end. Pharynx perforated by paired gill slits, which open, not to the exterior, but into a space formed by integumental folds, the peri-branchial cavity. This communicates with the exterior through a median opening, anterior to the arms, the atriopore.

The nervous system consists of a dorsal cord lying upon the notochord. It possesses a central lumen which enlarges a little at the anterior end. The sense organs are a set of cilia about the mouth, a median olfactory pit and a median pigment-fleck lying upon the anterior end of the nerve cord.

The reproductive glands are a series of pairs of follicles projecting into the peri-branchial cavity. The germ cells become free by rupture of the follicular walls, and later pass into the water through the mouth or the atriopore. The excretory organs consist of paired nephridia in the branchial region, which pass from the true coelom into the peri-branchial cavity.

Cladus IV. Vertebrata.

The main group of Chordata, characterized by the possession of a head and two pairs of lateral fins, which may become modified in various ways. The anterior portion of the central nervous system becomes enlarged to form a brain, which consists of five vesicles, and bears three pairs of sense-capsules, nose, eye and ear.

The notochord, which ends anteriorly at about the level of the ear-capsules, is cartilaginous, in some stage at least, but in the higher forms is reinforced by segmentally arranged osseous arches and other elements alternating with the mesodermic somites, which may entirely replace the original structure (*vertebral column.*) The brain and sense capsules are protected by a cranium, which consists primarily of two pairs of cartilaginous elements, to which may be added osseous tissues. Visceral arches always appear, but in the higher terrestrial forms, only show their primitive relationship in larval or embryonic life, and later undergo modification and reduction. The germ-glands are surface-gonads, i. e. localized portions of the coelomic mesoderm.

The excretory organs develop as paired nephridia, appearing in several series : these are consolidated to form kidneys, or utilized to convey the germ-cells.

Sub-cladus I. Cyclostomata.

Mouth circular, suctorial, without true jaws. Within the mouth are "teeth" of pure epidermic formation, and not homologous with the teeth of higher vertebrates. Paired limbs fail. The gill arches connect with each other and form a sort of basket-work in the pharyngeal region.

Sub-cladus II. Gnathostomata.

Mouth slit-like, transverse, furnished with jaws, which are originally modifications of the first visceral arch. The integument develops horny plates, formed from both epidermis and cutis, some of which become localized on the jaws as true teeth, others by fusion giving rise to the so-called "dermal" bones, which reinforce the skeleton. Paired limbs typically present. Appendicular skeleton develops, consisting of girdles about the body with which the skeleton of the free-limbs articulates.

Synopsis and Definitions of the Classes and Orders.

A. PROTOZOA.

Class I. RHIZOPODA . . . . Protozoa without a cell wall, moving by pseudopodia.

Order 1. Protoplasta . . . . Fresh water rhizopods found crawling over submerged objects, and on damp bog-moss, either naked or enclosed by a shell made of chitinous plates, sand grains, etc., with one large opening, out of which the animal extends its pseudopodia.

Sub-order 1. *Protoplasta lobosa* . . With thick, generally rounded pseudopodia. ★ *Amoeba, Difflugia.*

Sub-order 2. *Protoplasta filosa* . . With fine, thread-like pseudopodia. ★ *Euglypha.*

Order 2. Heliozoa . . . . Free-swimming fresh water Rhizopods, radiate in structure and often possessing a silicious skeleton, but without central capsule. ★ *Actinophrys, Clathrulina.*

Order 3. Radiolaria . . . . Marine Rhizopods, with radiate, silicious skeleton. Protoplasm divided by a central capsule into extra- and infra-capsular plasm. ★ *Polycystina.*

Order 4. Foraminifera . . . . Marine Rhizopods, with calcareous shell, perforated with very numerous minute openings (*foramina*) through which the animal projects its fine pseudopodia. ★ *Globigerina, etc., in deep-sea dredgings.*

(18)

B. METAZOA.

Division I. PROTAXONIA.

Type I. PORIFERA.

Class I. CALCISPONGIAE...........Spicules calcareous—structure of sponge comparatively simple.

Order 1. Asconida... ....Wall of sponge not folded—main cavity lined with digestive cells. ★ *Ascetta.*

Order 2. Syconida...............Wall of sponge forming oblong chambers opening directly into the main cavity. Digestive cells confined to the chambers. ★ *Sycandra.*

Order 3. Leuconida.............Wall of sponge with a very thick mesoderm, in which are spherical chambers lined with digestive endodermal cells Chambers connected with the exterior and with the main internal cavity by branching canals.

Class II. SILICISPONGIAE..........Spicules silicious, sometimes woven together like threads. ★ *Euplectella, Spongilla.*

Class III. KERATOSPONGIAE... ....No spicules, skeleton a mesh-work of woven threads, or wanting. ★ *Euspongia.*

Type II. CNIDARIA (*Coelenterata*).

Class I. HYDROZOA.... ....Polyp form [HYDRULA] without mesenteries, only in a few instances producing coral. Medusa form (*Craspedota*) generally small, with velum, 4-8 radiating canals, uncovered sense organs, generally a long manubrium and no genital pouches. Marginal lobes wanting. Gelatinous middle layer without cells.

Order 1. Anthydrae.............Sexual cells produced by polyp-form. No medusa generation (free or reduced).

Sub-order 1. *Hydrida*..........Simple, fresh-water polyps. Solitary individuals reproducing asexually in summer and sexually at the approach of winter. Do not form colonies. ★ *Hydra*.

Sub-order 2. *Hydrocorallinae*...Tropical marine forms. polymorphic, in colonies, producing a sort of coral. ★ *Millipora*.

Order 2. Hydro-medusae.........Sexual cells produced by medusa, which may be 1) a free form produced by a sessile or floating colony of polyps; 2) a reduced form borne on a polyp colony ; 3) a free-swimming form without polyp ancestors. All marine.

Sub-order 1. *Gymnoblastea-anthomedusae* .. Sessile colonies of polyps, the separate individuals of which are naked, and not protected by the perisarc. The medusae possess ocelli and genital organs in the manubrium. ★ *Pennaria* with reduced medusae.
{ ★*Syncoryne*=polyp } from one colony.
{ *Sarsia*=medusa   }

Sub-order 2. *Calyptoblastea-leptomedusae*....Sessile colonies of polyps, perisarc forming cups for the protection of the individuals. Medusae with otocysts and genital organs on radial canals. ★ *Campanularia* with free medusa. ★ *Sertularia* with reduced medusa.

Sub-order 3. *Trachymedusae*...Free-swimming medusa with direct development (no polyps). Velum, 4-6 radiating canals, and other characteristics of hydro-medusae. Evidently forms with suppressed polyp generation. [Primitive forms that have not developed the polyp form—BROOKS.] ★ *Gerygonia*, *Liriope*.

3

Sub-order 4. *Siphonophora*....Free-swimming colonies of polymorphic individuals representing both types (polyp and medusa) and characterized by the occurrence of a complex of several persons, known as a *Cormidium*. In some forms the medusae become free, ★ *Vellela*; in the majority the medusae are reduced, and develop germ cells (generally one egg in female) in the manubrium. ★ *Physophora, Physalia.*

Class II. Scyphozoa.............Polyp form [Scyphula] with an invaginated protostoma, forming oesophagus and secondary mouth, mesenteries present, extending between oesophagus and body-wall. Gastrocoele containing mesenterial filaments. Medusa form (*acraspeda*) large, without velum, with very many branching radiating canals, with square mouth and four genital pouches, with marginal lobes on sub-umbrellar side. Body cavity containing gastral filaments. Gelatinous middle layer cellular.

Order 1. Anthozoa.............Polyps of Scyphula type, without medusa generation. Mostly coral producers.

Sub-order 1. *Hexacoralla*......Mesenteries and other parts in sixes. Many forms produce coral, showing the arrangement of the mesenteries. ★ *Metridium, Fungia.*

Sub-order 2. *Octocoralla* (Alcyonaria).......Mesenteries and tentacles eight — coral consists of a horny axis coated with a polyparium containing sklerites. ★ *Renilla, Corallium.*

Sub-order 3. *Tetracoralla*......Parts in fours. All fossil. (Palaeozoic.)

Order 2. Scyphomedusae.......Large medusae of the above description (under *Scyphozoa*), which alternate with reduced polyp generation of the Scyphula type. ★ *Aurelia.*

Supplementary Class. PLANULOIDEA....Minute endoparasites without mouth or intestinal lumen, resembling the planula larva of Cnidarian polyps. There are two groups: the *Dicyemidae*, found in the nephridia of Cephalopods, and the *Orthonectidae*, in Ophiuridea, Turbellaria and Nemertea. They have been considered as *Mesozoa*, between Protozoa and Metazoa, but are more probably Cnidaria, reduced by parasitism.

Type III. CTENOPHORA.

Class I. TENTACULATA.............With two long, prehensile, thread-like tentacles. ★ *Pleurobrachia*.

Class II. NUDA ... .Tentacles wanting. ★ *Beroe*.

Division II. HETERAXONIA.

Type IV. ZYGONEURA.

Sub-type I. PROTONEPHRIDOZOA (*Scolecida*).

Cladus I. Platodes.

Class I. TURBELLARIA.............Small, free-swimming Platodes, of oval shape; body very contractile and covered with cilia.

Order 1. Rhabdocoela...........Intestine a straight, unbranched tube. Pharynx simple; small forms. ★ *Mesostomum*.

Order 2. Dendrocoela..........Intestine dendritic. Pharynx tubular. Larger forms, very flat.

Sub-order 1. *Triclada*.........Intestine with three main trunks—one anterior, and two posterior. ★*Planaria*.

Sub-order 2. *Polyclada*.........Intestine like foregoing, but with more than two posterior branches. ★ *Planocera*.

Class II. TREMATODES.......... .Parasitic Platodes. Body of adult not ciliated, but generally furnished with sucking-discs. Intestine much as in Triclada.

Order 1. Monogenea (Polystomeae)....Ecto-parasites upon gills, integument, bladder, etc., of aquatic vertebrates. Sucking-discs, three or more. Development direct. ★ *Polystomum*.

Order 2. Digenea (Distomeae)....Ento-parasites with never more than two sucking-discs. Development by heterogony, through forms known as Sporocyst, Redia, Cercaria, etc., which inhabit several hosts, the first being a pond snail, or some allied form.
★ *Phasciola*. (*Distomum*).

Class III. CESTODES.............. ...Ento-parasitic Platodes without intestine. Adult form generally composed of "links" (*Proglottids*), which may be considered reduplicated abdomens, or as attaining the value of individuals of a colony. Larval form (*Cysticercus*) provided with sucking-discs, and often loops. In the typical forms the cysticercus encysts itself in the muscles or internal organs of its first host, and develops into the chainform only when swallowed by some special animal, which thus serves as its second host. ★ *Taenia*.

Cladus II. Vermes.

Class I. ROTIFERA...... ....Minute aquatic forms, which may be considered as modified Trochozoa. The prae-oral band of cilia persists, often modified to form a pair of contractile organs resembling rotating wheels. Protocoelom spacious. Alimentary canal differentiated into a stomach, containing chitinous teeth, an intestine and a cloaca into which the protonephridia and reproductive ducts empty. Generally a jointed forked appendage or "foot," projecting posteriorly and serving for temporary attachment. Bisexual, the males rare, and much reduced.
★ *Brachionus. Rotifer*.

Class II. GASTROTRICHA ....Minute aquatic forms about the size of
large Infusoria, having an independent
origin from Trochozoa and modified in a
different manner from the above. They
are elongated, somewhat flattened, and
possess a double band of cilia on the ventral side, derived from the ventral stripe
of the Trochozoön. Ciliated zones fail,
body generally terminated in a fork.
★ *Ichthidium.*

Class III. ENDOPROCTA............Aquatic forms, mostly marine—probably
to be viewed as Trochozoa, modified by
sessile life. The larvae are free swimming, but soon become attached by a
stalk developed from the apical pole.
The prae-oral zone (Trochus) develops
into a crown of ciliated tentacles, within
which are mouth and anus, as well as
the genital and nephridial openings. The
apical ganglion of the larva disappears
and is functionally replaced by another,
situated between mouth and anus. The
group is generally referred to Bryozoa,
q. v. ★ *Cruatella, Loxosoma.*

Class IV. NEMATODES.............Long, cylindrical worms without cilia.
Body cavity present. Sub-cuticula with
four longitudinal thickenings, dorsal, ventral and lateral, between which lie four
bands of longitudinal muscles. Protonephridia forming two lateral canals en-
cased in the lateral sub-cuticular thickenings. Sack gonads long and tubular,
often contorted. Ventral nervous system
an oesophageal ring, from which issue
longitudinal nerves, of which two (dorsal and ventral) lie in the sub-cuticular
ridges. Bisexual with a single exception.
Development direct. Free-swimming and

parasitic forms. ★ *Trichina, Anguillula* (vinegar-eel). *Ascaris* (pin-worm). *Gordius* (hair-snake).

Class V. ACANTHOCEPHALI........A single family of worms. externally similar to last, but with an anterior extensile proboscis beset with hooks, and no intestinal canal. The body wall develops two solid masses (*Lemnisci*) which project into the protocoelom. The reproductive organs open posteriorly. Development by metamorphosis. Larval stage in crustaceans and insects, adult in intestine of vertebrates. ★ *Echinorhynchus.*

SUPPLEMENTARY CLASS—NEMERTINI.

Flattened forms, often very long, externally covered with cilia. At anterior end a long retractile proboscis. Protocoelom filled with parenchyma. Alimentary canal complete with anal orifice. Nephridia and gonads repeated in pairs along the body. Complicated blood and nervous systems. Bisexual. Many develop by metamorphosis and possess a curious larval form, the "Pilidium," which is similar to a Protrochula.

The relations of this group are very uncertain. Many characteristics would place them near the Platodes, but the complete alimentary canal, the blood system and the segmentally arranged nephridia and gonads would exclude them from these and place them nearer the Annelides. It seems at present more probable that they are Protonephridozoa, and that the metameric character of some of the internal organs is due to a reduplication of protosomatic elements, rather than to the formation of a metasoma (cf. the reduplication of the abdomen in *Taenia*).

Order 1. Hoplonemertini........Proboscis armed with bristles.

Order 2. Schizonemertini........Proboscis unarmed, head nearly divided by deep, longitudinal fissures.

Order 3. Palaeonemertini........Proboscis unarmed. Head nearly entire.

Sub-type II. METANEPHRIDOZOA (*Aposcolecida*).

Cladus III. Articulata.

Sub-cladus I. Annelida.

Class I. ARCHIANNELIDES..........Small marine forms of very simple organization. No external segmentation. Metameres all alike : bristles, cilia and parapodia fail. Development with a metomorphosis, in which occurs the most typical trochophore. ★ *Polygordius*.

Class II. CHAETOPODA ....Typical Annelids, with well marked external segments, corresponding to the internal metamerism—segments furnished with paired groups of chitinous bristles.

Order 1. Polychaeta ....Bristles conspicuous, situated on raised lâteral portions, parapodia. Head generally present. Development by a metamorphosis, usually with a Trochophore larva.

Sub-order 1. P. *errantia*......Free-swimming, active, predaceous. ★ *Nereis*.

Sub-order 2. P. *sedentaria*.....Live in tubes built of sand, mud, bits of shell, etc., subsist upon vegetable substances. ★ *Amphitrite, Serpula*.

Order 2. Oligochaeta............Bristles very small, sunk in hollows along the sides ; no parapodia ; no distinct head; hermaphroditic, development direct.

Sub-order 1. O. *limicola*......Aquatic, in mud of swamps. ★ *Nais*.

Sub-order 2. O. *terricola*......In damp earth. ★ *Lumbricus*.

Class III. HIRUDINEA ....Aquatic hermaphroditic ecto-parasites : segments without bristles ; move by terminal, adhesive suckers. External and internal segments do not correspond.

Order 1. Rhynchobdellidae......Pharynx extensile, forming a sort of proboscis. ★ *Clepsine*.

Order 2. Gnathobdellidae. ......Pharynx not extensile, with three longitudinal ridges which are often toothed.
★ *Hirudo.*

-- -------- --

SUPPLEMENTARY CLASSES.

Class CHAETIFERA, formerly taken with Sipunculoidea to form class Gephyrea. Here belong a very few marine worms, *Echiurus, Bonellia,* etc., which
show affinities to the Chaetopoda. They are segmented only as larvae, but
possess paired nephridia and a system of blood vessels similar to that of
Annelids.

Class SIPUNCULOIDEA includes a very few forms of marine worms, bearing
some slight affinity to Annelids. They were formerly united with Chaetifera
to form the class Gephyrea. They are cylindrical, unsegmented forms without bristles, and possess a retractile proboscis. ★ *Sipunculus, Phascolosoma.*

Class CHAETOGNATHA. This includes one form, Sagitta, a small, transparent, unsegmented worm found on the surface of the ocean. The body is flat
and possesses lateral fire-like extensions. The mouth is armed with jaws beset
with sharp hooks. Hermaphroditic. It develops lateral coelom sacks, which
appear to have the value of a metacoelom. It is placed by some with the
Nematodes.

Sub-cladus II. Arthropoda.

Class 1. CRUSTACEA................Aquatic Arthropoda, a few being secondarily adapted to a terrestrial life. Res-
piration either through the general surface of the integument or by localized
thin portions of the same, in the form of
evaginated plates or structures, almost
always placed in some relation to the
appendages—and known as "gills." Appendages typically composed of two
branches, which may be modified beyond
recognition. Development from a Nauplias larva, which is suppressed in the
higher forms. Female generally provided with a brood-sack for the care of
the young.

Sub-class I. ENTOMOSTRAKA........Small. often minute Crustacea with a variable number of segments (not 20). Abdomen generally without appendages. Nauplius larva almost universal. Parthenogenesis frequent.

Order 1. Branchiopoda............With flat, leaf-like legs bearing gill-sacks. Body generally enclosed in an integumental duplicature in the form of lateral shells or a dorsal shield.

Sub-order 1. *Phyllopoda*......Body plainly segmented. Numerous pairs of legs (10-40). ★ *Branchipus*.

Sub-order 2. *Cladocera*.......Body enclosed in a shell with two lateral valves—head free ; second pair of antennae enormously developed, and used as oars. 4-6 pairs of legs. Gill-sacks may fail. ★ *Daphnia*.

Order 2. Copepoda...............Body generally elongated and plainly segmented. without integumental duplicature. 4-5 pairs of flattened. two-branched legs used as oars. Eggs in two lateral pedicelled sacks. Many forms reduced by parasitism.

Sub-order 1. *Eucopepoda*......Free-swimming forms with typical characteristics. ★ *Cyclops*.

Sub-order 2. *Copepoda parasitica*......Parasites upon gills of fish or in internal organs. Often with a free-swimming stage, especially in male. In later life may be reduced to a shapeless sack. Antennae may be modified to form hooks. ★ *Lernaea*.

Sub-order 3. *Branchiura*.......Flattened oval forms. attached to the sides of fish and subsisting upon slime. ★ *Argulus*.

Order 3. Ostracoda ....Body, including head, enclosed in a bivalve shell. with hinge and adductor muscle. seven pairs of appendages. of which only 2 (or 3) may be reckoned as legs. ★ *Cypris*.

Order 4. Cirripedia.... ....Sessile forms, enclosed in an inverted position in a calcareous 2-valved shell. Generally six pairs of 2-branched legs, modified to form attenuated many jointed cirri. Larval stage a free-swimming nauplias which soon becomes fixed by the first antennae. This after passing through a so-called "Cypris" stage, develops into the adult.

Sub-order 1. *Lepadoidea*. .....Forms with stalk and flexible valves—skeleton mainly composed of scuta, terga and carina. ★ *Lepas*.

Sub-order 2. *Balanoidea*.......Forms without stalk, skeleton reinforced by lateral pieces, which, with carina and rostrum, form a calcareous tube. ★ *Balanus*.

Sub-order 3. *Rhizocephala*... .Degenerate parasites upon crabs (*Brachyura*). Body consists of a sack, from which grow countless root-like threads (=the stalks, morphologically) which penetrate the flesh of the host. Recognizable as Cirripedia only in larval life. ★ *Sacculina*.

Sub-class II. MALACOSTRAKA......Generally large forms with a constant number of segments (20) consisting of a head with five, a thorax with eight and an abdomen of seven segments. The first two portions are often fused to form a cephalo-thorax of thirteen segments. Only the first six abdominal segments bear appendages, and of these the last pair is generally modified and united with the terminal segment to form a caudal appendage. The paired reproductive orifices of the male are found upon the last thoracic segment, and those of the female upon the third from the last. Development sometimes direct—generally with a metamorphosis. Nauplias larvae appear only in a few primitive forms.

Legion 1. Phyllocarida (*Leptostraka*)....This group, mostly fossil, contains but a single pelagic form, valuable as a link between Entomostraka and Malakostraka. A small integumental duplicature covers the head and thorax. The abdomen is 8-jointed, thus departing from the constant number found in Malacostraka. The feet are like those of Phyllopods. ★ *Nebalia.*

Legion 2. Arthrostraka (Edriophthalmata)......Sessile-eyed Malacostraka, with free thoracic segments, and without carapace. Thoracic appendages distributed as one maxilliped and seven legs. Brood cavity borne between thoracic legs.

Order 1. Amphipoda.... ....Compressed forms, body generally bent into a curve. Gills upon the thoracic legs. Abdomen with six pairs of legs, of which the first three are used in swimming, and the last three form a springing organ. In a few forms the abdomen is reduced. ★ *Gammarus.*

Order 2. Isopoda ....Depressed forms with seven free thoracic segments. The abdominal appendages are in the form of flattened plates and protect the gills. ★ *Porcellio.*

Legion 3. Thoracostraka (Podophthalmata)....Eyes situated upon movable stalks. Carapace involving all or nearly all the thoracic rings. Brood cavity upon the ventral side of the flexible abdomen and protected by the sixth abdominal appendage and the terminal segment (telson).

Order 1. Cumacea........ ......Carapace small, involving only 3-4 thoracic segments, two pairs of maxillipeds and six pairs of legs. Abdomen of female without appendages; of male with 3-5. ★ *Diastylis.*

SUPPLEMENT TO THE CRUSTACEA.

TRILOBITEA.

An important group of fossil forms, presenting some superficial resemblance
to the Isopods. They are oval and flattened in form, and possess a dorsal carapace, divided by two longitudinal grooves into three areas, one median
(*rhachis*) and two lateral (*pleurae*). The carapace consists of a cephalo-
thorax, and a variable number of segments, of which the anterior ones are
free-moving, and termed the " thorax," followed by several fused ones, the
" abdomen " or *pygidium*.   ·

The cephalo thorax consists of a central piece (*glabellum*), and two lateral
pieces, the fixed and movable cheeks, between which are a pair of compound
eyes.

The ventral side is almost unknown. Probably each thoracic segment has a pair of crustacean-like legs, above which there may have been a pair of gills, protected by the overhanging edge of the carapace. Recently (1894) some finely preserved specimens have been discovered showing one pair of long antennae.

The Trilobites appear to have close affinity to the branchiate Arachnoids, as well as to the genuine Crustacea.

Class II. ARACHNOIDEA........ ....Head and thorax fused into a single piece, the cephalo-thorax, bearing six pairs of appendages, of which one is prae-oral. These may all be used as legs, or one or more pieces may be chelate or toothed and serve as mandibles or weapons of defense. In the lower forms the abdomen is elongated and segmented and may bear appendages, but in the higher forms it is consolidated and may be fused with the cephalo-thorax. Respiratory organs originally lamellate gills, developed as adjuncts of the abdominal appendages. In the air breathing forms they may be modified and reduced in number, or even replaced by a sort of tracheal system, not homologous with that of insects.

Sub-class I. ARACHNOIDEA BRANCHIATA....Mainly fossil forms, all marine, gills lamellate, one pair of eyes (" trilobite ") in side of cephalo-thorax, and one pair of small ones anterior to these near the middle line. Coxal joints of the legs, or of some of them forming spiny plates used in mastication.

Order 1. Gigantostraka... ......Fossil forms with long extended abdomen, which may terminate in a spine.
★ *Pterygotus.*

Order 2.   Xiphosura (Limuloidea)...... Mostly fossil, with three surviving species—Abdomen consolidated in recent forms and bearing six plate-like appendages, of which the last five bear lamellate gills. A long terminal spine at end of abdomen. The larvae pass through a " trilobite " stage, showing three well marked longitudinal areas—and without the terminal spine.   ★ *Limulus.*

Sub-class II.   ARACHNOIDEA TRACHEATA...Mostly terrestrial forms—breathing either by gill plates hanging from the roof of pneumatic chambers, or by tracheal tubes. Both sorts communicate with the exterior by paired stigmata situated on the ventral side of the abdomen. 1-6 pair of eyes—generally simple. First two pairs of appendages employed as mouth parts, often chelate. Last four serve as legs.

Order 1.   Scorpionidea........... Cephalo-thorax of one piece, broad prae-abdomen of seven segments and narrow, elongated post-abdomen of five, ending in a poisonous spine. Mandibles chelate. Second pair of appendages enormously developed and chelate. Four pairs of respiratory chambers.   ★ *Scorpio.*

Order 2.   Pseudoscorpionidea..... Cephalo-thorax of one piece. Abdomen 11-jointed, broad and flat, without attenuated portion or sting. First and second pairs of appendages as in Scorpionidea. Breath by tracheal tubes, which open by two pairs of stigmata on second and third abdominal segments. Spinnerets on the second abdominal segment. Small animals found under bark, etc.
   ★ *Chelifer.*

Order 3. Pedipalpi...............Cephalo-thorax of one piece. Abdomen broad and depressed, 11-12 jointed, without antennuated post-abdomen, but in some cases ending in a long filiform process. Mandibles with single claw. Second pair of appendages large and chelate. Third pair feeler-like, extended like whip lashes. Two pairs of respiratory chambers on second and third abdominal segments. ★ *Phrynus.*

Order 4. Solpugidea ....Head distinct, thorax of three segments, abdomen cylindrical, 10-jointed—mandibles chelate. Second pair of appendages leg-like, not chelate. Tracheal respiratory system with stigmata on the first thoracic segment and second and third abdominal segments. Nocturnal animals found in sandy parts of the tropics—bite poisonous. ★ *Solpuga.*

Order 5. Phalangida............Entire body a shortened oval. Reduced abdomen closely applied to cephalo-thorax, but distinct and consisting of six segments. Mandibles chelate. Legs very long and attennated. Tracheal tubes with one pair of stigmata between thorax and abdomen. ★ *Phalangium.*

Order 6. Arachnida... ....Abdomen without segments, swollen and attached to cephalo-thorax by a stalk. Mandibles ending in a simple claw with poison gland. Second pair of appendages leg-like—modified in male. 2-3 pairs of spinnerets at end of abdomen. 1-2 pairs of respiratory chambers situated on abdomen. These may be also connected with a system of tracheal tubes. ★ *Epeira.*

Order 7. Acarina.................Abdomen fused with cephalo-thorax.
Body unsegmented. Appendages about
mouth often modified to form a sucking
tube. Respiration by tracheal tubes or
merely through integument in smaller
forms—many parasitic.
★ *Ixodes, Tyroglyphus.*

SUPPLEMENTARY GROUPS.

Pantapoda.......................Extremely attenuated marine forms with
long slender legs and body of about the
same diameter. Cephalo-thorax of six
segments, of which the first are fused,
possessing a beak at the anterior end.
Abdomen much reduced and sac-like.
Seven pairs of legs, containing the reproductive organs and diverticula of the
stomach. Respiratory organs fail.
★ *Nymphon.*

Tardigrada.......... ....Minute fresh water forms, with four
pairs of short legs bearing little hooks.
Hermaphroditic—without heart or organs
of respiration. ★ *Macrobiotus.*

Pentastomidea (Linguatulina).....Parasites in the lungs and nasal cavities
of reptiles and mammals. Long, flat-
tened worms, resembling Taeniae, but
with Arachnoid development. Legs reduced to two pairs of hooks about the
mouth. ★ *Pentastomum.*

Class III. TRACHEATA ANTENNATA.. .Head always distinct, never fused
with thorax, one pair of antennae, breathe
by system of tracheal tubes opening by
stigmata.

Sub-class I. ONYCHOPHORA ..... This includes one genus of widely separated species, intermediate between Annelids and Myriapods. The body is worm-like, each segment bearing a pair of limbs with indistinct articulations, thus resembling parapodia. Paired nephridia of Annelid type, opening at the base of the feet. A richly branched tracheal system, opening by irregularly dispersed stigmata. ★ *Peripatus*.

Sub-class II. MYRIAPODA .... Head with antennae, mandibles, and two pairs of maxillae. No differentiation of thorax. Each segment bears a pair of legs, alike in shape and size.

Order 1. Chilopoda............ ..... Mainly depressed in form. One pair of legs to each segment. Mandibles well developed, fitted for predaceous life. First pair of legs transformed to a pair of jaws furnished with poison glands. Reproductive opening at posterior end of body. ★ *Lithobius, Scolopendra*.

Order 2. Diplopoda....... ...A double pair of legs to each segment. Maxillae united to form a complex underlip, the gnathochilarium. Reproductive opening at base of second pair of legs. ★ *Iulus*.

Sub-class. HEXAPODA (Insecta).....Body divided into three distinct regions, or segment-complexes, head, thorax and abdomen. Mouth parts typically of three component parts, the mandibles, and the first and second maxillae, which may become strongly modified in adaptation to very varied life habits. Thorax of three segments, each one with a pair of legs. The last two segments in higher insects bear each a pair of wings formed by an integumental duplicature and directly derivable from " tracheal gills," a modification originally for aquatic breathing and still retained in a few aquatic larval forms. The higher insects have gained secondary larval forms and thus develop by a metamorphosis.

Order 1. Thysaneura... ....Minute wingless forms, with biting mouth parts, found in decayed wood and damp earth. They are the most primitive insects and have never developed wings. Some show rudiments of of abdominal legs. Ametabolic, i. e. development direct. ★ *Podura*.

Order 2. Pseudoneuroptera......Mouth parts biting, wings all alike, transparent, delicate, with lace-like venation. This group was formerly united with the Neuroptera, but has an active pupa (hemimetabolism). ★ *Libellula*, *Ephemera*.

Order 3. Orthoptera............Mouth parts biting—upper wings parchment-like, generally narrow—under wings membraneous and often folded. Development hemimetabolic i. e. pupa active. ★ *Caloptenus*.

Order 4. Neuroptera.... ......Mouth parts biting—somewhat modified in *Phryganidae*. Wings as in Order 2. Development holometabolic i. e. pupa—quiescent. ★ *Phryganea*.

Order 5. Coleoptera............Mouth parts biting—upper wings (elytra) forming hard shields for the protection of the membraneous lower ones. Holometabolic. Thorax free and quite distinct from the posterior portion, thus dividing the body into three regions, head, pro-thorax, and meso- and meta-thorax + abdomen. Holometabolic. ★ *Carabus*.

Order 6. Rhynchota (Hemiptera)..Mouth parts modified to form a straight jointed beak, which lies between the coxal joints of the legs. Wings either membraneous and alike, or with the outer diagonal half of the upper wings pergamenteous. Many forms wingless. Hemimetabolic. ★ *Coreus, Cicada.*

Order 7. Lepidoptera ....Mouth parts a double coiled proboscis, formed by the first maxillae. Wings alike in texture, membraneous, covered with minute colored scales. Holometabolic. ★ *Papilio, Sphinx.*

Order 8. Diptera...............Mouth parts variously modified, sucking, piercing or lapping—never biting. Fore wings membraneous, hind wings reduced to minute knobs—the so-called balancers. Holometabolic. ★ *Culex, Musca.*

Order 9. Hymenoptera. ....Mouth parts biting—or biting and lapping. Wings membraneous, alike in texture, but hind pair reduced in size. Body generally much constricted between thorax and abdomen. Ovipositor of male generally accompanied by organs for sawing, boring or digging, and in some a venomous sting. Holometabolic. ★ *Vespa, Formica.*

Cladus II. Mollusca.

Class 1. AMPHINEURA ....A small group of bilaterally symmetrical marine molluscs with very primitive characteristics. They have a nerve ring around the mouth, from which pass two lateral and two ventral nerve cords connected by transverse commissures.

Order 1. Solenogastres.......... Worm-like forms without mantle or shell, but with calcareous spicules imbedded in the cuticula. Some possess a ciliated ventral furrow, in the bottom of which lies a rudimentary foot. In others furrow and foot fail. ★ *Proneomenia.*

Order 2. Chitones .... Depressed forms with a dorsal shell composed of eight transverse pieces. Ctenidia in two longitudinal rows between foot and edge of mantle. Foot large, with oval, flattened creeping surface. ★ *Chiton.*

Class II. LAMELLIBRANCHIATA...... Bilaterally symmetrical Mollusca with four (in a few cases two) plate-like gills, and with two lateral shells generally united dorsally. They are without a distinct head and lack a "lingual ribbon" or radula. The foot is compressed and never forms a creeping disc.

Order 1. Siphoniata.............. With two posterior siphons, separate or fused. Edges of mantle often joined.

Sub-order 1. *Sinupalliata*...... Siphons long and contractile. Mantle line with sinus. ★ *Mya, Venus, Solen.*

Sub-order 2. *Integripalliata*.... Siphons short, not contractile. No pallial sinus. ★ *Cyclas.*

Order 2. Asiphonia.......... .... Siphons absent.

Sub-order 1. *Homomyaria*..... Anterior and posterior adductor muscles about equal. ★ *Unio.*

Sub-order 2. *Heteromyaria*.... Anterior adductor very small. ★ *Mytilus.*

Sub-order 3. *Monomyaria*...... Anterior adductor wanting. ★ *Ostrea, Pecten.*

Class III. SCAPHOPODA ....Bilaterally-symmetrical forms with body cavity greatly elongated in a dorso-ventral direction. Mantle and shell tubular and somewhat curved, with a smaller dorsal and a larger ventral opening. Ctenidia fail. Foot elongated and conical. A single family of marine forms. ★ *Dentalium.*

Class IV. GASTEROPODA.... ......Molluscs with a head, foot and visceral sack. The first two are bilateral, the third is almost invariably unsymmetrical, the pallial complex, with its organs being developed upon one side only (usually the right). The visceral sack is generally contained in a spirally twisted shell, wound usually toward the right about a central axis, and capable of receiving the other parts when contracted. The mouth is furnished with a radula; the foot generally forms a creeping disc.

Order 1. Prosobranchiata. ......Shell present, ctenidium anterior to heart ; foot a creeping disc ; bisexual. ★ *Olira, Conus, Cypraea, Strombus.*

Order 2. Heteropoda........... Shell small or wanting. Gill as in previous order. Anterior part of foot compressed, forming a sort of keel. Bisexual. This order includes a very few nearly transparent forms, which swim on the surface of the ocean. They may be considered as Prosobranchiata adapted to a pelagic life. ★ *Carinaria.*

Order 3. Pulmonata.............Land and fresh water snails, breathing by plexus of blood vessels, which lie in a respiratory chamber communicating with the exterior, and placed anterior to the heart. Ctenidia fail. Shell generally present. Hermaphroditic. ★ *Helix, Limnaea.*

Order 4.   Opisthobranchiata......Shell delicate or wanting. Respiration seldom by ctenidia, often by secondary or adaptive gills, or through the integument. Gills, when present, lie behind the heart. Back of naked forms often ornamented with simple or dendritic papillae. Hermaphroditic. ★ *Bulla, Eolis,*

Order 5.   Pteropoda..............Shell fragile or wanting, foot developed into a pair of wing-like expansions. Hermaphroditic. A small group of forms which swim at night upon the surface of the ocean—often referred to preceding order, from which they have undoubtedly been derived. ★ *Cymbuliopsis.*

Class V.   Cephalopoda............Body bilaterally symmetrical, extended in a dorso-ventral direction and flattened antero-posteriorily so that the anterior aspect seems dorsal, the posterior ventral, the dorsal posterior, etc. The greatly modified foot forms a series of tentacles about the mouth and a funnel or infundibulum behind it. The head is large and distinct, with two large prominent eyes. Mouth provided with a pair of chitinous jaws. Shell large and chambered, or reduced or even internal.

Order 1.   Tetrabranchiata........Gills four, mouth surrounded by numerous unarmed tentacles. Ink bag fails. A heavy external shell convoluted and divided into chambers—the animal being in the terminal and largest one. ★ *Nautilus (only living form).*

Order 2.   Dibranchiata ....Gills two. Eight (or ten) arms around the mouth, covered with cup-shaped sucking discs. Ink bag present. Shell internal (a very fragile external shell in *Argonauta*). ★ *Octopus, Loligo, Sepia.*

The present Cephalods are the few degenerate descendants of a very large and abundant group, which filled the seas in Palaeozoic and Mesozoic times. They possessed, originally, well developed shells, divided into chambers : some shells being straight, others spirally coiled. The orders of NAUTILOIDEA and AMMONOIDEA were Tetrabranchs, the BELEMNOIDEA Dibranchs.

Cladus III. Molluscoidea.

Class I. BRYOZOA (Polyzoa) ...... Minute forms, usually colonial. At anterior end a ridge, the lophophore, which bears ciliated tentacles. Anus situated outside the lophophore.

Order 1. Phoronidea ... . Worm-like forms, enclosed in leathery tubes. Larva free-swimming—the " *Actinotrocha.*" Similar to the trochophore —one genus. *Phoronis.*

Order 2. Ectoprocta .... Typical forms, nearly always forming a colony, which resembles an alga. Each animal is enclosed in a transparent cell, from which it may extend its tentacles, and into which it may entirely withdraw. A few fresh water forms, the rest marine. ★ *Bugula, Plumatella.*

Class II. BRACHIOPODA .... All marine, depressed in form, with dorsal and ventral shells, which are symmetrical, but unequal. Mouth situated between two spiral ciliated arms, which lie coiled up in the shell—a large fossil order. Few living.

Order 1. Ecardines .... Shell without hinge. ★ *Lingula.*

Order 2. Testicardines ... . Shell with hinge, usually calcareous loops to support arms. ★ *Terebratulina.*

Type V. AMBULACRALIA.

Cladus. Echinodermata.

Class 1. HOLOTHUROIDEA...... ...Adult creeping or sessile, oval or vermiform, covered by a leathery integument in which minute calcareous spicules, plates, etc., lies imbedded. Around the mouth is a crown of fringed retractile tentacles. Ambulacral feet, either in rows, or irregularly disposed, or wanting. Larval form the *Auricularia*—reduced in some cases.

Order 1. Pedata..................Ambulacral feet present. ★ *Pentacta, Thyone.*

Order 2. Apoda..................Ambulacral feet wanting. ★ *Synapta.*

Class II. CRINOIDEA..............Mostly sessile, flower-like forms; with stem, calyx and many-branching arms covered with pinnulae. Very few living forms — many fossil. Free-swimming larval form oval, with ciliated bands. ★ *Pentacrinus.*

[The following fossil classes belong here. They are often included among Crinoids:

Class. CYSTIDEA.....................Mostly with arms, calyx plates irregular.

Class. BLASTOIDEA................Without arms, calyx plates regular.]

Class III. ASTEROIDEA............Adult star-shaped to pentagonal, with exo-skeleton in the form of a rough network, studded with fixed spines. Ambulacral feet in grooves on oval side.

Order 1. Asteridea..............No clear distinction between mouth and arms. Larval form the *Bipinnaria* and *Brachiolaria.* ★ *Asterias.*

Order 2. Ophiuridea..............Disc and arms distinct, the latter serpentine and very brittle. Larval form a *Pluteus*, similar to that of Echinoidea. ★ *Ophiopolis.*

Class IV. ECHINOIDEA............Adult spheroidal, oval or disc-shaped, with exo-skeleton composed of solid calcareous plates arranged in meridional or radial rows. The surface of this shell is beset with spines which rotate upon tubercles. Rows of ambulacral feet project from foramina in shell.

Sub-class 1. PALECHINIDA.........Shell with more than twenty rows of plates—all fossil.

Sub-class II. EUCHINIDA..........Shell with twenty rows of plates, ten ambulacral and ten interambulacral.

Order 1. Regularia...:..........Mouth and anus in the center of their respective surfaces. ★ *Strongylocentrotus.*

Order 2. Clypeastroidea.........Mouth central, anus eccentric. ★ *Echinarachnius.*

Order 3. Spatangoidea...........Mouth and anus both eccentric. ★ *Spatangus.*

Type VI. CHORDATA.

Cladus I. Hemichorda.

This Cladus was founded for a single genus of worm-like marine forms, found in mud-flats. At the anterior end is a long, flexible proboscis, with which the animal pushes its way through the mud ; at the base of this is a narrow zone, the collar, at the upper ventral edge of which the mouth is situated. This is followed by a long worm-like body, showing paired gill-slits on its ventral aspect. The main details of its structure have been given under the description of the Cladus, q. v. ★ *Balanoglossus.*

[*Cephalodiscus* and *Rhabdopleura*, two sessile forms classed as Order, Pterobranchia, under Bryozoa, have been found to resemble Balanoglossus. In young buds of the former a division into proboscis, collar and body may be seen. There are also a single pair of gill-slits, and a dorsal diverticulum of the intestine (= notochord ?) lying under the dorsal nervous system. In *Rhabdopleura* no gill-slits have been detected, but in other respects the structure is similar to *Cephalodiscus.* HEMICHORDA may thus be represented by two classes : Class I, ENTEROPNEUSTA, including the different species of *Balanoglossus*, and Class II, PTEROBRANCHIA, including the two forms under consideration.]

Cladus II. Urochorda.

Class I. PERENNICHORDATA........Free-swimming forms, like the larvae of higher tunicates. They possess a long tail provided with a skeletal axis, the notochord. Pharynx with a single pair of gill-slits. No definite mantle, but a gelatinous envelope. ★ *Appendicularia.*

Class II. CADUCICHORDATA........Tail and notochord present only in larval life.

Sub-class I. ASCIDIACEA...........Body sack-like. Pharyngeal wall forming a sort of lattice-work. Excurrent and incurrent orifices generally approximated.

Order 1. Ascidiae...............Sessile forms, either solitary (*monascidiae*), or colonial (*synascidiae*) and arranged in generally stellate groups, known as *coenobia*. ★ *Boltenia, Botryllus.*

Order 2. Pyrosomiae...........Free-swimming, transparent colonies of cylindrical or cone-shaped forms. Incurrent openings upon the exterior, cloacal openings in the interior. ★ *Pyrosoma.*

Sub-class II. THALIACEA...........Free-swimming, transparent cask-shaped forms. Pharynx with two rows of small gill-slits, or a single pair of large gill-slits. Oval and cloacal openings at opposite poles—often with alternation of generations.

Order 1. Doliolidae .. ..Two rows of gill-slits. Muscle-bands in the form of closed rings. Mantle thin—generative cycle including one sexual and two asexual generations. ★ *Doliolum.*

Order 2. Salpidae ... .A single pair of gill-slits. Muscle-bands not as complete rings. Mantle thick. Alternation of generations simple, solitary asexual individuals alternating with a chain-like series of sexual forms. ★ *Salpa.*

Cladus III.   Cephalochorda.

This group includes but two closely allied genera. The anatomical details are given in the definition of the Cladus—q. v. ★ *Amphioxus.*

Cladus IV. Vertebrata.

Sub-cladus I. Cyclostomata.

This Sub-cladus includes two groups which may have the value of families. For general anatomical details see the definition of the main group.

Family I. *Petromyzontidae* .....Nasal cavity a median blind-sack. Mouth
(Lamprey eels.) without tentacles. They attach them-
selves by their circular mouth to the
sides of fishes and suck their blood.
Fresh water and marine forms.
★ *Petromyzon.*

Family 2. *Myxinidae* ....Nasal sack provided with an inner pala-
(Hag-fish.) tal opening. Mouth with tentacles.
Habits more parasitic than the former.
They push their way even into the body-
cavity of other fish and consume their
viscera. All marine. ★ *Myxine.*

Sub-cladus II. Gnathostomata.

Division 1. ANAMNIA...........Amnion and other embryonic membranes
fail. Breathe by gills during at least a
portion of their life. Mesonephros functions as permanent kidney.

Class I. PISCES....................Aquatic Anamnia, with branchial respiration. Paired limbs in the form of fins
(*Ichthyopterygium*). Integument generally produces scales, which contain bony
material produced by the corium. Head
with one auricle and one ventricle.

Order 1. Selachii (Elasmobranchii)..Skeleton cartilaginous, scales placoid.
Tail heterocercal. 5-7 separate gill open-
ings without operculum. ★ *Squalus.*

Order 2. Ganoidei..............Skeleton more or less reinforced by bone.
Scales overlaid with enamel and typically
rhomboid in form (ganoid). Tail heterocercal. Gill-slits with operculum.

48 INVERTEBRATE ZOÖLOGY.

Sub-order 1. *Chondrostei*......Notochord an unsegmented cartilaginous rod, upon which the arches and other vertebral elements rest. Integument with rows of bony plates, or naked. ★ *Accipenser. Polyodon.*

Sub-order 2. *Holostei*.........Notochord with bony constrictions, corresponding to the vertebral centra. Skull with membrane bones. ★ *Lepidosteus.*

Order 3. Teleostei..............Skeleton osseous. Scales cycloid or ctenoid. Tail homocercal. Gills provided with an operculum.

Sub-order 1. *Physostomi*.......Swimming bladder provided with a pneumatic duct. Ventral fins abdominal. Fin rays soft. Almost all the fresh water fish belong here. ★ *Salmo.*

Sub-order 2. *Physoclisti*.......Swimming bladder without connection with alimentary tract, often wanting.

Group 1. *Anacanthini*.......Fin rays weak. Ventral fins anterior to pectoral. ★ *Gadus.*

Group. 2. *Acanthopteri*......Some fin rays spinous, at least of the dorsal fin. Inferior pharyngeal bones separate. The largest group of Teleosts. ★ *Scomber.*

Group 3. *Pharyngognathi*....Some fin rays spinous. Inferior pharyngeal bones separate. ★ *Labrus.*

Group 4. *Plectognathi*.......A small group of very compact forms. Maxilla and prae-maxilla immovably joined with skull. ★ *Diodon.*

Group 5. *Lophobranchii*.....Gills tufted, eggs carried by male in a brood sack. Body covered by large plates. ★ *Hippocampus.*

Class II. DIPNOI.............. ...Skeleton mainly cartilaginous. Notochord persistent. Paired fins with a central skeletal axis and with or without lateral rays. Median fin continuous around tail. Swimming-bladder functions as lung. Gills with operculum. In one species small external gills. A very small, but isolated group, including 3-4 species. These represent two orders.

Order 1. Monopneumones........Lung (=swimming-bladder) single. ★ *Ceratodus.*

Order 2. Dipneumones..........A pair of lungs present. ★ *Protopterus.*

Class III. AMPHIBIA................Skin naked, very glandular and slimy. Gills, external and internal, present at some stage, generally transitory. Lungs usually present, with larynx and trachea. Paired limbs of the hand-form (*cheiropterygium*) with normally five digits, often reduced in number. Two occipital condyles.

Order 1. Urodela...............Body elongated. Tail persistent.

Sub-order 1. *Perennibranchiata*....With external gill-bushes and median fin in caudal region. ★ *Necturus.*

Sub-order 2. *Derotremata*......No external gills, but with a persistent gill-slit. ★ *Menopoma.*

Sub-order 3. *Salamandrida*....External gills and gill-slits embryonic or larval. Adult breathe by lungs, or by pharyngeal and integumental respiration, the lungs failing. ★ *Salamandra, Desmognathus.*

Order 2. Gymnophiona..........Serpentine subterranean forms without tail, limbs or gills. Integument with minute scales sunken in pits. ★ *Cœcilia.*

Order 3. Anoura...............Compact, cephalized forms, with tail present only in larval life. Hind legs enormously developed and used for leaping and swimming. No gills when adult. ★ *Rana, Bufo.*

Division II. AMNIOTA ....Embryo furnished with amnion and allantois. Mesonephros and its duct possess an excretory function only in the embryo, being replaced in adult life by the definite kidney (*metanephros*). The mesonephros then disappears, except certain portions, which are utilized by the reproductive system.

Class IV. REPTILIA ... .Integument covered with scales, horns, and other structures of epidermic formation. Glands confined to a definite locality (femoral glands of lizards) or wanting. Gills never developed, embryonal gill-arches modified to subserve other functions. A single occipital condyle.

Sub-class 1. PLAGIOTREMATA... ....Cloacal opening transverse, behind which, in the male, are paired organs of copulation. Body uniformly covered by delicate scales, which are cast off yearly often as a single piece. Habits terrestrial and arboreal. Quadratum movably articulated with skull.

Order 1. Lacertilia.............Four well-developed limbs in the typical gano, and sternum or shoulder girdle present in forms with reduction of limbs.

Sub-order 1. *Fissilinguia*......Tongue long and slender, extensile, forked at the end. ★ *Lacerta*.

Sub-order 2. *Brevilinguia*......Tongue short and thick, thin and notched at tip. Limbs sometimes fail. ★ *Anguis*.

Sub-order 3. *Crassilinguia*.....Tongue short and thick, rounded at tip, not extensile. ★ *Iguana, Gecko*.

Sub-order 4. *Vermilinguia*.....Tongue very long, vermiform, thickened at tip. ★ *Chameleo*.

Sub-order 5. *Annulata*.... ....No limbs or eyelids—Epidermis divided into oblong fields by longitudinal and transverse forms. ★ *Amphisbaena*.

Order 2. Ophidia............ ....Body very much attenuated, limbs fail. no rudiment of shoulder girdle or sternum. A single lung developed (right). Other paired organs placed the one behind the other. ★ *Crotalus, Python*.

[A single species of a very ancient type of lizard occurs in New Zealand, *Sphenodon (Hattena) punctata*. The quadrate is immovable, ventral ribs and abdominal sternum are present, the vertebrae are amphicoelous. It is referred to the Order Rhynchocephalia.]

Sub-class II. HYDROSAURIA........Cloacal opening oval in shape, its longer axis longitudinal. A single organ of copulation in the male, anterior to the cloaca. Scales large and irregular, often reinforced by bony plates, which may coalesce to form dorsal and ventral shields. Mainly aquatic in habits. Quadratum immovably attached to skull.

Order 1. Chelonia.. ....Body enclosed by dorsal and ventral shields, formed partly by elements of the endo-skeleton and partly from the integument. Teeth replaced by horny beak, with sharp cutting edge. ★ *Chrysemys, Chelone*.

Order 2. Crocodilia.............Body elongated, covered by large plates, which do not coalesce. Thoracic and abdominal sterna present, connected by ventral and dorsal ribs. Teeth large, in alveoli. ★ *Alligator*.

[Our knowledge of Reptilia is greatly increased by the discovery of several fossil classes and orders. The PTEROSAURIA were allied to the Lacertilia, and possessed membranous expansions of the integument of the arms and fingers, by which they could fly. The PLEISIOSAURIA and ICHTHYOSAURIA were hydrosaurs, the former somewhat resembling turtles, the latter crocodiles. The DINOSAURIA included some enormous terrestrial forms with massive skeletons, *Brontosaurus, Iguanodon*. Other smaller Dinosaurs may have been the precursors of birds. They walked mainly upon their hind feet, possessed pneumatic cavities in their bones, and showed many other avian characteristics. An important form is *Compsognathus* (one specimen at Munich) which shows affinity to *Archaeopteryx* (see introduction to Aves)].

Class V. AVES..................Warm-blooded Vertebrates. showing
                                many reptilian characters, but differing
                                from them in the possession of feathers.
                                These are epidermic structures closely
                                related to scales. The anterior limb is
                                modified to form a wing, generally giving
                                the animal the power of flight. Quad-
                                rate bone movably articulated with the
                                skull. A single occipital condyle.

[Some have recommended the fusion of Reptilia and Aves into a single class, *Sauropsida*,
which is warranted not only on anatomical grounds, but by the possession of intermediate
fossil forms. It seems, however, more practical to retain them as separate classes, which if
we take living forms alone into consideration, is easy to do. The following fossil groups
may be given, preceding the living forms. They should have the value of Sub-classes:

SAURURA...........................Jaws containing teeth, tail elongated, containing
                                a large number of vertebrae, with a pair of con-
                                tour feathers to each. Digits of the wing not
                                coalesced, three being armed with claws.
                                Archaeopteryx.

ODONTORNITHES..... ....Bird-like forms similar to recent birds, but with
                                teeth in both jaws, inserted either in separate
                                alveoli, or in a common groove. There are two
                                groups of these birds, one having a keelless ster-
                                num allied to the Ratitae, the other with a keeled
                                sternum and allied to the Carinatae.
                                Hesperornis, Ichthyornis.]

Sub-class 1. RATITAE..... ......Breast bone flat. Clavicles not united to
                                form a furcula. Feathers down-like or
                                plume-like. Running birds with small
                                or rudimentary wings. Cannot fly.
                                ★ *Struthio.*

Sub-class II. CARINATAE ....Breast bone keeled. Furcula generally
                                present. Contour feathers on wings and
                                tail.

Order 1. Gallinacei..............Feet stout, for scratching. Hind toe on
                                a higher level than the others. Edge of
                                upper beak shuts over lower beak. Sec-
                                ondary sexual characters common, in
                                form of combs, wattles, spurs, etc.
                                ★ *Gallus, Perdix.*

Order 2. Columbini............Legs short, hind toe on a level with the others. Edge of upper beak in contact with that of lower. At base of upper beak two outgrowths covering the nostrils. ★ *Columba.*

Order 3. Natatores... ....Aquatic birds with oily feathers and short webbed feet. ★ *Anser, Larus.*

Order 4. Grallatores.............Wading birds with very long slender legs, and toes without webs. Often the lower half of the tibia is free from feathers. Neck and bill very long and slender. ★ *Grus, Scolopax.*

Order 5. Scansores.............Climbing birds. Inner forward toe reversible, giving the foot two toes in front and two behind. Several groups of birds belong here, not closely related. ★ *Psittacus, Picus.*

Order 6. Passeres...............The most numerous group. Feet fitted for perching. Two groups, the *Oscines,* or singing birds, which have developed a special organ, the Syrinx, at the forking of the bronchi; and the *Clamatores,* without this. ★ *Fringilla, Cypselus.*

Order 7. Raptores ....Birds of prey. Toes furnished with hooked claws or talons. Point of upper beak sharp and talon-like, projecting over the lower one. ★ *Aquila, Strix.*

Class VI. MAMMALIA..............Warm-blooded vertebrates. Body clothed with hair; young nourished by milk, a secretion of integumental glands. Quadrate in middle ear, two occipital condyles.

Sub-class I. MONOTREMATA........Low oviparous mammalia with reptilian characteristics. The young, immature when hatched, are brooded either in a nest or in a brood pouch temporarily developed. No localized mammary glands;

5

the " milk " is a secretion of perspiratory glands, richly developed in a certain area, the mammary pocket, and thus not strictly homologous with the milk of other mammals. Alimentary canal, urethra, and reproductive ducts open into a common cloaca. ★ *Ornithorhynchus, Echidna.*

Sub-class II. MARSUPIALIA... ..... Viviparous mammals without placenta. The young are born in a very immature condition and brooded in an external abdominal pouch, the *marsupium.* Two lateral uteri imperfectly united. Division between anus and sinus uro-genitalis internal and indistinct.

Order 1. Zoophagi.............Carnivorous marsupials with pointed teeth and well developed canines. ★ *Didelphys.*

Order 2. Phytophagi.............Herbivorous marsupials with flat teeth and reduction of canines.   ★ *Macropus.*

Sub-class III. PLACENTALIA....... Young nourished in uterus of mother by a capillary mass, the placenta, which adheres to the uterine wall and is connected with the embryo by the umbilical cord. Anus separated from the uro-genital sinus by a *perinaeum.*

Order 1. Edentata.............Teeth either wanting or in condition of retrogressive metamorphosis. Incisors and canines generally fail. Large number of sacral vertebrae.   ★ *Dasypus.*

Order 2. Cetacea...............Aquatic forms with naked skin, provided with hair in embryo. Hind limbs fail externally. Rudiments under skin. Mammary glands in folds on the sides of the vagina. External nares in top of head connecting directly with a tubular prolongation of the larynx.   ★ *Balaena.*

Order 3. Sirenia................Aquatic forms with only the anterior
limbs developed. Teeth often fail; mo-
lars, when present, resemble those of ungulates. Sparsely hairy. A single pair
of pectoral mammary glands.
★ *Manatus.*

Order 4. Ungulata.............Herbivorous mammals with flattened
molars. Toes often reduced in number,
tipped with flattened hoofs.

Sub-order 1. *Perissodactyla*....Ungulates with an odd number of toes,
five, three or one, the middle one being
the best developed. Prae-molars equal
the molars in size. Integment often very
thick. ★ *Rhinoceros, Equus.*

Sub-order 2. *Artiodactyla*....Ungulates with an even number of toes,
of which two, (3 and 4) are the best de-
veloped, resulting typically in the cloven
hoof. The prae-molars, 3-4, are smaller
than the molars. A part of them are
ruminants. ★ *Bos, Hippopotamus.*

Order 5. Proboscidia............A group allied to the Ungulates, but
with always five stout toes furnished
with hoofs, making a ponderous rounded
foot. The snout is enormously prolonged
forming a muscular proboscis tipped with
a sensitive finger-like process. One pair
of incisors, which develop into mam-
moth tusks. In recent forms it is those
of the upper jaw. (Lower incisors in
Dinotherium.) ★ *Elephas.*

Order 6. Rodentia............A group of small animals with teeth fitted for gnawing. One pair of incisors
in each jaw develop into sharp cutting
chisels. Canines fail. The molars are
fitted with transverse ridges for cutting.
★ *Mus, Sciurus.*

Order 7. Insectivora............Small insect-eating mammals with all the teeth prolonged into sharp points. Canines small or wanting. ★ *Talpa.*

Order 8. Carnivora.............Beasts of prey with sharp well developed canines and pointed molars. The toes are armed with claws, which may become sharp and retractile. There are two groups, one terrestrial, *Fissipedia*, with toes separate, and one aquatic, with toes strongly webbed, forming paddles— *Pinnipedia.* ★ *Felis. Phoca.*

Ooder 9. Cheiroptera...........The only flying mammals. Toes of anterior limb exceedingly attenuated, thus forming a frame work for a thin leathery web, which also includes the hind limb and tail. Thumb of fore limb and all the hind toes free. Teeth pointed as in Insectivora. In many respects similar to the apes, as a discoidal placenta and a single pair of pectoral mammary glands. ★ *Vespertilio.*

Order 10. Prosimii.............A group of animals closely allied to the apes, but of a generally lower structure. Appendages hand-like, with opposing thumbs, but with a double uterus, and a diffuse placenta. The nails are developed into claws.

Order 11. Primates.............Toes with flat nails. Appendages more or less hand-like, and generally fitted with opposing thumbs for grasping. One pair of pectoral mammary glands. Placenta discoidal.

Sub-order 1. *Platyrrhini*.......Nose flattened, nostrils separated by a broad septum, so that their orifices look outward. Confined to the new world. ★ *Cebus.*

Sub-order 2. *Catarrhini*......Internasal septum thin. Nostrils look forwards and downwards. ★ *Cercopithecus, Gorilla.*

Sub-order 3. *Anthropini*......Nose as in Sub-order 2. Thumb more movable — great toe less so. Hair reduced. Erect position normal. ★ *Homo.*

[There are three main races of men, which are thought by some to have the value of species. Their characteristics and distribution are as follows :

1. *Negroes*.........................Hair curly, oval in cross-section, skin densely pigmented, nose flat, lips projecting—includes : *Papuans, Hottentots, Kaffirs* and *Sudanese.*

2. *Mongolians*.......................Hair straight, circular in section, skin brown or yellow, cheek bones high, lips thin—includes :— *Malays, Eskimos, Mongols, American Indians.*

3. *Caucasians*.......................Hair wavy or curly, circular in section, skin white, nose and lips thin -includes :—*Semites, Indo-Europeans, Nubians, Dravidas.*]

Provisional phylogenetic table, to express the relationships of the main groups as given in the Synopsis. Hypothetical ancestral forms are under-scored twice, larval forms once.

II. Laboratory Practicum.

LABORATORY PRACTICUM.

PROTOZOA.

I. PROTOPLASTA. ("Rhizopods" in restricted sense.)

These are minute globules of protoplasm. i. e. "cells" without cell wall. Some are naked, others protected by a shell. Occur in enormous quantities in slime at bottom of ponds and in swamps, particularly in the moisture collected upon bog-moss (*Sphagnum*).

1. Squeeze out a drop of water from a tuft of Sphagnum ; let it fall in the center of a clean slide : cover and examine 100[1]. cf. Leidy, Plates. Study several drops in this way and identify. If a form is interesting, bring it to the center of the field and adjust higher power 300-600[1].

2. Collect a few slimy leaves and sticks from a stagnant pond or ditch. Place in a shallow dish with water from the same place. [This is best transported in a closed can.] Put a drop of this water upon a slide, mix with it the scrapings from a slimy leaf, cover and examine as above. (*Continue the above investigations [1 and 2] until you have identified three of the following forms :—Amoeba, Difflugia, Euglypha, Nebela, Arcella, Hyalosphenia, Quadrula, Cyphoderia. Make a drawing of each of the three forms, noting the structure and the character of the shell when present.*)

3. Study of living forms. Life is indicated by the movement of the protoplasm. Amoebae will always be found alive, as so minute a drop of dead protoplasm would disintegrate at once. On the other hand, the shells of the other forms are very enduring and are often empty. In such shells a deceptive appearance of life may be caused by swarms of minute infusoria, etc.. which sometimes inhabit them.

3. *a.* If the form is an Amoeba, draw several successive shapes. Note the granular endoplasm and clear ectoplasm. Are the pseudopodia pointed or blunt ? [cf. Leidy for different species.] Study the particles of food. What do Amoebae eat ? Can you find a circular pinkish object that appears and disappears ? This is the contractile vacuole. Can you see the nucleus? This is also circular and like the surrounding protoplasm, but more refractive.

3. *b.* In case of most shelled forms, the protoplasm cannot be well observed through the thick walls of the shell. In these, watch the large aperture of the shell and look for pseudopodia which may project from it, often for some distance (nearly the longer diameter of the shell). These may be finger-like (*Difflugia*) or exceedingly delicate and filamentous (*Euglypha*). In some delicate shells (Hyalosphenia, Cyphoderia) the interior protoplasm may be seen, connected with the shell by tapering threads. Encysted forms may be observed without pseudopodia, and the protoplasm in the form of a sphere within the shell.

3. *c.* Keep watch for reproducing Rhizopods. In the shelled forms the new one grows from the aperture of the shell. When fully formed one sees a pair of shells placed mouth to mouth.

4. (This experiment is not an essential one, but valuable for comparison). The " white blood corpuscles" or Leucocytes are amoeboid cells with nucleus and pseudopodia. They occur in the blood and lymph as well as in the saliva and in the alimentary canal. They may be studied for comparison. Immerse a fine needle in 100% Alc. to cleanse it ; when dry again, prick the finger with it, and bring a drop of blood upon a slide. Dilute with clear saliva, free from bubbles, and cover. (The saliva is the best fluid for diluting the blood. It keeps the form of the red corpuscles and is itself rich in leucocytes.) Search for leucocytes. When found, examine for several minutes 300-600ᵈ and observe the changes of form. These are rendered extremely slow by the lowering of the temperature and soon stop. (cf. the difference between 98.6° F., the temperature of the body and that of the microscopic stage.) The movements are better seen by use of the " warm stage," an apparatus which heats the preparation by means of an alcohol lamp, or by using the blood of some animal of lower normal temperature, as the frog.

II. HELIOZOA.

These are radiate, free-swimming forms, found in clear water, captured best by the towing net. They are thus difficult to find and never at hand when wanted. Bear them in mind, however, and be ready to recognize the first one you meet in subsequent study of pond water. When one is found, study and draw it, and if possible identify it by Leidy.

III. FORAMINIFERA

These are marine and cover with their shells the bottom of the ocean over vast areas, and often at great depths. In such places the "ooze" dredged from the bottom is mainly composed of Foraminifera shells.

5. Pour out a little of the dry ooze into a watch crystal and examine with dissecting microscope. Find as many different shapes as possible. A common form is the *Globigerina* form (several genera) a series of spheres of different sizes, irregularly heaped together. Another common shape is *Rotalia*, and allied forms : flat and consisting of several chambers arranged in a spiral.

6. Place a few shells on a slide under a dissecting microscope and add a drop of HCl. What happens? Of what material do the shells consist?

7. [The experimental part of this will furnish material for the entire class, and need not be performed by the student.] Boil gently a quantity of ooze in a test-tube of 70%. This is to expel air from the separate chambers. Allow it to settle and then decant off the 70% and replace with 95%. In the same way add 100% and turpentine, after which the preparation is brought into a shallow dish. A stay of several hours in each liquid is advisable. To prepare a permanent mount from this, pipette a drop upon a clean slide, drain off the excess of turpentine and mount in Balsam. In this preparation observe the numerous pores, or foramina, through which in life the very numerous filamentous pseudopodia are projected.

IV. INFUSORIA.

8. Place a handful of leaves, flower-stems, etc., with a pinch of sugar in a beaker of water and allow it to stand a week or two, until the water is of a greenish yellow color. Such water forms an "infusion," and will be found filled with *Infusoria*.

[In this as well as in subsequent experiments requiring time, make them as stated, and pass on to the next, returning to these at the proper time.] The Infusoria obtained by this experiment are mostly oval flattened forms, the type being called *Paramoecium*.

8. *a.* Bring a drop upon a slide, cover and examine. Study them in full motion as well as you can. By a little skill you can keep a moving form under continual observation, keeping the left hand on the slide, the right one on the focal screw and the eye at the microscope. The left hand follows the movements in an horizontal plane, and the right controls the up and down motion. This will only succeed with low powers, for the higher powers limit the field and exaggerate the velocity.

8. *b.* Mount a drop in a loose tuft of cotton. The animals will be confined between the threads and are easier to study. Watch their motions and find out if possible by what organs they move.

8. *c.* Look over several drops, if necessary, until you find a double one like the figure 8. Watch its movements and seek to explain the phenomenon.

8. *d. Staining intra vitam.* If Bismark Brown be added to a drop containing living forms, and withdrawn after a few minutes, the nuclei will be found stained, the animals remaining alive. A colony may be kept alive several days in a glass of weak Bismark Brown solution, with the same result.

9. Search among several different localities, i. e. different ponds, swamps, ditches, etc., as well as the different laboratory aquaria, for other forms of Infusoria. They may generally be distinguished from other forms by their rapid movement. Other Protozoa and most one-celled plants move very slowly or not at all. The few sessile forms show their Infusorial nature by the possession of cilia and by the movements of the parts of the body. Forms which are especially to be noted are the following :—

a. The trumpet form, *Stentor*, a very large green Infusorium.

b. The " Bell animalcule." The commonest forms are : (1) *Vorticella*, attached by spiral stalks to leaves and stems ; (2) *Epistylis*, a large branching colony with rigid stems, often found upon the backs of snails ; (3) *Cothurnia*, In little cups without stalks, found upon antennae of *Cyclops*.

c. Volvox—Colonies of flagellate infusoria arranged in the form of beautiful green globes. Look in these for sexual cells and daughter colonies.

d. Monads—Simple flagellate forms, often green, distinguished by the long flagellum. Many of these forms are the flagellate stage (swarm-spores) of fresh water algae.

10. Kill a frog, open the rectum and mount in a drop of water a bit of the slime from its walls. It will generally contain parasitic infusoria. *Balantidium* similar in form to *Paramoecium*.

V. GREGARINIDA.

These are vermiform unicellular forms, found as parasites in animals of every class. The largest ones are visible to the naked eye, but the majority are microscopic, often living within a single cell. *Monocystis* is a common form infesting earth-worms. The adult, often visible to the naked eye, is found in the body cavity, lying along the wall of the alimentary canal. The encysted form, filled with pseudonavicellae, occurs in the spermatic vesicles and is found in one out of about three or four individuals.

11. Open an earth-worm along the mid-dorsal line, from the anterior end as far as a thickened girdle of lighter color, the clitellum (about the anterior fourth). Pin it out in a dissecting pan and examine with the lens. The spermatic vesicles consist of three pairs of very noticeable organs somewhat kidney shaped and covering the alimentary canal dorsally. In infected specimens, minute white flecks may generally be found scattered over the surface of these organs, and often on the body wall, and other organs in the vicinity. When such a specimen is found, remove to a slide a bit of the infected portion, tease it out a little and press with a cover glass, but not too hard. With 50ᵈ the white flecks may be seen to be cysts filled with pseudonavicellae. These vary greatly in size and number contained. Select a small and clear one and observe separate pseudonavicellae with a high power. Crush a cyst by pressing a needle upon the cover glass *while under observation,* 50–100ᵈ. The individuals may thus be isolated and studied separately.

VI. BACTERIA.

12. Collect in a water drop a very slight amount of the scum floating upon the top of stagnant water, or water containing decomposing material. Scum from macerating bones is especially good. Cover and observe with high powers. Focus high—i. e. for the upper surface of the layer of water. Look for extremely minute sticks and dots lying in all directions. These are stick forms, *Bacillus* and *Bacterium.* The screw form, *Spirillum,* is often present. These may all be in active motion or in a quiescent condition.

13. Make a permanent mount of the above, as follows :

(1) Spread a very thin layer of the material containing Bacteria upon a cover-glass.

(2) Set aside and allow to dry.

(3) Hold glass in the fingers, preparation side uppermost, and pass quickly through an alcohol flame 2–3 times. This coagulates the thin pellicle of albumen surrounding each individual and thus fastens it to the glass.

(4) Lay the glass down and place upon it a drop of methyl violet, fuchsin, or methyl blue, and let it stand 3–5 minutes.

(5) Take the glass in forceps and wash off with stream from wash bottle.

(6) Lay it down, preparation side up, and *allow to dry.*

(7) Place then a drop of Canada Balsam in the center of a clean slide, *invert* the cover glass and place it upon the drop.

14. Scrape the inside of the check very slightly with a dull knife and mount a little of the substance obtained in a drop of clear saliva—100^d. The field will be filled with large flat cells that look like thin membranes. Find one that is isolated from the rest and well spread out. Place this under 300-600^d. In the center will be seen its nucleus, a clear oval mass. Surrounding this will be seen numerous round dots, highly refractive. These are ball-bacteria. Make a permanent mount of this as given above [13].

PORIFERA.

Type I.—An Ascon Sponge.

15. Place an entire specimen, or group of several, in a watch crystal of 70$_\zeta$ and study under dissecting microscope. Notice manner of growth, main body with osculum, and lateral buds. Have any of these oscula? Cut open with scissors and notice main cavity. Have the lateral buds cavities? Do they communicate with the main chamber? The texture of the sponge resembles felt. Cause of this? Draw the group with above details expressed.

16. Place a portion of the wall (single thickness) in a solid watch glass of Borax Carmine. Cover it and leave an hour. [While waiting, conduct experiments 17-19 and then return.] Wash out in 70$_\zeta$, 5-10 minutes. Then five minutes each in 95$_\zeta$—100$_\zeta$—turpentine—and place in a drop of Balsam on a slide, *inner wall uppermost,* cover and examine. Most superficially come the endodermal cells, which in life are like collared monads. By focussing a little deeper the spicules come in view, and deeper yet the irregular cells of the mesoderm. Certain of these are very large—spherical and deeply stained. These are either eggs (one large cell) or spermatozoa (a mass of cells). Draw these details.

17. Select another portion of the wall, spread it out on a dry slide, inner wall uppermost, and allow to dry. When perfectly dry, examine for spicules, 50-60^d.

18. Place a few pieces in a watch crystal of K-O-H, and heat gently over an alcohol lamp. Pipette the residuum upon a slide, cover and examine. 50-60^d. Compare with dried specimen. Express the results of the last two experiments by drawings showing (1) Isolated spicules, (2) Relation of spicules to each other. Include in the examination the spicules about the osculum.

19. Mount a small piece temporarily in 70ℊ. Add a drop of HCl at edge of cover and watch result through microscope. Of what are the spicules composed?

Type II.—A Sycon Sponge.

20. Study a specimen (or small group) as in 1. Do not cut open with scissors, but use a razor or a sharp scalpel. Divide the sponge longitudinally, and make a few cross sections of one longitudinal half. Examine all these in the watch crystal. If you succeed in obtaining a very thin cross section, make a temporary mount of it and examine, 60ᵈ. Note the thick walls, the narrow central chamber and the pores in its walls. What do these signify? Does an examination of the sections of the walls solve this problem? Have these pores any regular order? Note the palisade of large spicules about the osculum.

21. Repeat 4 for spicules.

22. *Study of prepared sections. Method and introduction.* The walls of this specimen are too thick to be mounted whole as in 2, and we must resort to microtome sections. The two most useful are longitudinal and transverse sections.

[For this purpose specimens are prepared as follows :
(1) Borax Carmine, 36 hours and thus stained *in toto.*
(2) Washed out in " acidulated alcohol," i. e. 70ℊ + a few drops of HCl, until bright scarlet, 3-5 minutes.
(3) 95 ℊ—3-6 hours.
(4) 100ℊ—3-6 hours.
(5) Turpentine—3-6 hours.
(6) In paraffine, in oven, 1-2 hours and imbedded.]

Consider the form of the specimen and answer the following :—Will one sponge yield more transverse or more longitudinal sections? Which set will vary more among themselves? Will all longitudinal sections include the lumen? Longitudinal sections lying in the plane of a radius of a transverse section are termed *radial,* those not including a radius, but merely parallel to a tangential plane, are termed *tangential.* How many of each sort? Are all radial sections the same? Are all tangential sections? If the longitudinal axis of a specimen curves, how will that complicate the sections? What sort of specimens should be selected for sectioning?

23. Study of a transverse section. Note the radial canals, characteristic of the type, the outlines of which form a symmetrical rosette. The canals are lined with endoderm cells of the same sort as those lining the main chambers of the Ascon. Between these are the spicules—broken by the knife in cutting. Can you detect any regular arrangement of these? Notice the long spicules forming an external cone capping each canal. What is the shape of a canal? Is it possible to tell from cross-sections alone? Why do some of the canals appear more or less filled up with cells?

24. Study of a longitudinal section. How do the canals appear in this? Are they differently shaped in the middle and at the ends of the section. Explain this.

After the above study, construct a series of diagrams representing the complete structure of the sponge, giving the relation of canals and partitions, arrangement of spicules, etc. These are not to be copies of the sections. Sections are generally imperfect, cut obliquely, torn, etc. A diagram should be without imperfections and should combine the result of all different modes of observation, and thus be a graphic method of representing one's knowledge. In constructing these diagrams let each student select the view and devices which seem to her the best means of expressing the structure. Thus for instance : a perfect cross-section, with whole spicules, occupying such positions as will illustrate their plan of arrangement. The effect is heightened if delicate colors be used to represent the different parts [cf. Dendy's plates in Quarterly Journal for 1893]. A radial chamber, drawn as a solid body, and covered with spicules in regular arrangement, would express the shape well and explain the different planes seen in the sections.

Additional study of sponges :—

25. Spicules may be isolated from a piece of silicious sponge by teasing or by K-O-H [cf. 4]. The shapes may be studied and their composition tested by HCl. This should not be added to a specimen in K-O-H, as the two reagents will counteract each other. If the fresh water sponge, *Spongilla*, be macerated in K-O-H, especially in the case of specimens collected during the autumn, gemmules may be isolated as well as the spicules.

26. *Skeleton of a horn sponge.* With a sharp razor make several very thin sections of a perfectly dry piece of bathing sponge and drop them into a watch crystal of turpentine for five minutes. Select the thinnest one and transfer it to a slide, spreading it out in a drop of turpentine. This structure, which is the part we use, is merely the skeleton of the sponge animal, the cellular portions having been removed by maceration or boiling. This skeleton is neither silicious nor calcareous, but composed of closely interwoven threads of *chitin* (same material as horn, silk, etc). Draw a small area of this under a high power to show the relation of the threads to each other. This specimen may be permanently mounted by removing the cover, wiping away the excess of turpentine and adding a drop of Canada Balsam.

Type III.—Hydra.

The Hydra is the only fresh water representative of the group. It is found in quiet pools, clinging to submerged vegetation. It is tubular or vase-shaped with 6–8 long thread-like tentacles hanging from the free end. There are two principle forms distinguished by their colors—the brown one, *Hydra fusca*, and the green one, *Hydra viridis*.

27. *Field work.* Search for specimens in stagnant water abounding in aquatic plants. Hydras average $\frac{1}{4}$ inch in length and may be sometimes seen attached to submerged leaves and stems, their long tentacles streaming in the water. They are easily seen in an aquarium, but to find them in a natural pool is much more difficult. The best plan is to collect a handful of aquatic plants, place them in a jar filled with clear water and hold it up to the light. The observation should not be too hasty, for hydras are extremely sensitive and contract at once when so roughly handled. One should therefore wait a few minutes and allow them to expand. Repeat this test several times and in different localities.

28. *Study of external form.* Place a hydra in a watch crystal of water and observe with a dissecting microscope. Make outline drawings of several shapes and positions. Place under the low power of the compound microscope, and study the structure of the tentacles, body wall, etc. Make a sketch, using an outline (one of those just made), and fill in what you see. It is better to draw one tentacle and a little portion of the body-wall minutely and leave the rest in outline.

29. *Study of life habits.* Keep two or three specimens for several weeks in a large beaker, placed where it can be frequently watched. Feed with minute fresh water Crustacea (Cyclops, etc). The results of this experiment depend largely upon luck and continual observation. If fortunate, one may observe (1) the prey caught by the tentacles and conveyed to the mouth lying between them ; (2) the growth of new individuals from the sides of the older ones ; (3) the subsequent separation of the offspring as an independent hydra, etc.

30. [*This difficult experiment is placed here only as one which may be done, and should be attempted only by patient investigators, who are prepared for repeated failures.*] Trembly, a French zoölogist, in 1744, cut living hydras in two and found that under favorable conditions each piece would grow again into a perfect hydra. He even cut one into several pieces and found the larger portions were capable of becoming complete animals.

This experiment is easy to attempt, but results are hard to obtain. Cut several hydras in different ways—transverse, longitudinally, etc., and keep the portions in a beaker of clear water. Scissors may be used for this, but a sharp razor is better.

31. *Study of a cross-section.* To kill a hydra in an expanded condition, place in a watch crystal with a very little water, and suddenly deluge the animal while expanded with a test tube full of hot alcoholic corrosive sublimate [70% + $HgCl_2$]. Allow it to stand a few minutes until cool and then place in a large amount of pure 70%, which should be changed once after a few hours. Such a specimen may be treated as the Sycon Sponge in 22 and sectioned. In the section, notice the two layers of cells, ectoderm and endoderm, separated by a definite line, the supporting lamella. Sections of different animals and of the same animal at different planes, show different points, as follows:

a) *Ova*, large cells with amoeboid processes, crowded in between the two layers, but belonging to the ectoderm. Sections at about the middle of the body show this.

b) *Testes*, a conical mass of cells, belonging to the ectodermic layer, forming a protuberance. Mostly in the upper portion of the vase-shaped body.

c) *Cross-section of a tentacle.* Notice the few huge endoderm cells which form a lining and enclose a minute central lumen continuous with the gastrocoele.

d) *Nettle cells, (nematocysts).* These may be found anywhere, even in endoderm, but are most common at the ends of the tentacles. They are large, oval capsules, semi-transparent, and much more refractive than the surrounding tissue.

Type IV.—Campanularia.

32. Make preliminary examination of material in a dish of 70%. Notice the two forms of polyp, each enclosed in a transparent cup. Select and cut off with scissors one or two good branches for mounting. These should include specimens of both sorts of polyps; and, if possible, in different stages

6

and conditions—expanded, contracted, buds, empty cups, etc. Branches, the members of which lie approximately in one plane, are the best for this purpose.

33. Mark out on a sheet of white paper a series of rings the size of watch crystals, and designate as follows :

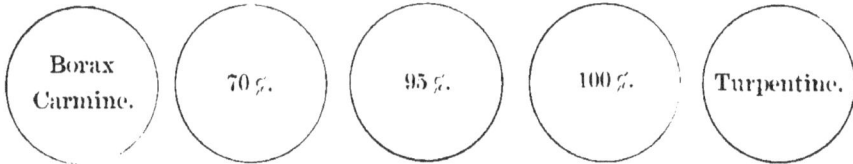

| Borax Carmine. | 70 ℅. | 95 ℅. | 100 ℅. | Turpentine. |

Place watch crystals on these and fill with the corresponding liquids. Use a solid watch-glass for the 100℅ and keep it covered. Drop the branches selected into the borax carmine and let them stay 5-10 minutes, until deeply stained. Then transfer successively to the different watch crystals, 4-5 minutes in each. Use a needle for this, but do not break or prick the specimens. From the turpentine spread out the branch on a slide in a small drop of balsam. Cover and examine. The living protoplasmic portion is colored pink and consists of the polyps and irregular connecting stems between them, running through the centre of the branches. The whole is invested by a transparent chitinous skeleton, the *perisarc*, which encloses the stems and expands into a cup about each polyp. Study details as follows :—

a) A feeding polyp or *Hydranth*. The body is divided by a constriction into a terminal portion (*manubrium*) which bears the tentacles, and an expanded basal portion. How many tentacles are there? Is the number constant? What can be seen upon these with the high power? The basal portion is hollow, containing the gastrocoele, or gastro-vascular cavity. Around the hydranth is an expanded cup, the *hydrotheca*. Can a polyp entirely retract into this? In this species the hydrotheca has the shape of a bell (*campanula*), hence the name. Just below the polyp the perisarc is divided into rings or joints. What is the probable use of them? Is their number constant? Does it bear any relation to the age or size of the polyp?

b) A reproductive polyp or *Blastostyle*. This is merely the stalk which bears round medusa-buds ; it may be considered a reduced polyp, elongated in shape, and without mouth or tentacles. It is surrounded by a cup, the *Gonotheca*. Compare with a hydrotheca.

c) *Medusa-buds*. These are produced as buds from the sides of the Blastostyle. Which of these are the most mature? In what direction does growth progress? Study the oldest of these buds. Can you find traces of medusa structure? [cf. 35].

34. Express by drawing the results obtained. As such may be suggested : 1) a sketch of an entire branch, only a few times enlarged, drawn from dissecting microscope. 2) An enlarged figure of each sort of polyp. Make this a perfect and symmetrical diagram, taking the details from several if necessary. Thus one specimen may have the best manubrium, another may show good tentacles, another a perfect hydrotheca, etc. 3) Drawing of details as seen with a high power. Thus a portion of an enlarged tentacle, showing nettle cells, or a bit of body wall seen in "optical section" i. e. focused to show the thickness of a lateral wall ; the specimen is thus cut by the focal plane of the microscope. Such detailed drawings as these may be drawn beside (2) and connected by dotted lines with the corresponding part.

35. *Study of a free medusa.* These are caught in the open ocean with a tow net and may often be found in preserved specimens of tow. In this species they are of the size of pin-heads, disc-shaped and bordered by a fringe of tentacles. Search for them as follows : Place a watch crystal of 70% on stage of dissecting microscope. Pipette into this a drop of thick tow. This dilution will serve to isolate the forms, which may then be sorted over with a dissecting needle, looking continually through the lens. In case of fresh living tow, use sea-water instead of 70% for the dilution. When one is found, it may be stained and mounted as in 33, but handled with a pipette instead of a needle. In this, reduce to a minimum the amount of liquid transferred with the specimen. In case a specimen becomes lost, place the crystal upon the stage of the dissecting microscope and search as at first. In a well mounted specimen there may be seen :—(1) the manubrium, (2) four radiating canals, (3) four genital masses *lying upon the canals*, (4) the tentacles with *otocysts* at the base of some of them.

OTHER STUDIES OF HYDROZOAN FORMS.

36. *Pennaria* is a common form and shows many important differences from the foregoing. Prepare and study in the same way and note, 1) the perisarc does not grow up over the polyps, but stops short at their bases, leaving them unprotected. 2) The medusa-buds do not grow upon a blastostyle, but at the bases of the ordinary polyps, which are of but one kind, all hydranths. 3) These buds do not become free, but remain as reduced *gonophores*, producing the genital products in the place of origin. The gonophores are bisexual and produce eggs or spermatazoa. These unite in the water and develop into a polyp which may develop asexually into another colony. 4) There are also minor differences in the shape and arrangement of the tentacles, the nettle cells, the stem, etc.

37. *Thamnocnidia* (or *Clava*). These are solitary polyps, although generally associated in clumps. Each polyp bears at its base very numerous medusa-buds or gonophores in racemose clusters. The male gonophores liberate the spermatozoa, which escape into the water ; but the female gonophores retain the eggs which are fertilized and develop *in situ*, and leave it in the form of minute polyps, which pass out backwards, drawing their tentacles after them. Cross-fertilization is insured because a given polyp produces gonophores of only one sex ; the polyp, although asexual, is thus often termed male or female. *Thamnocnidia* is too large for mounting entire. It may be studied in a watch crystal, and separate gonophore-clusters mounted.

Clava may be mounted by cutting little strips of cardboard, soaking them a few minutes in turpentine and placing them in the balsam with the specimen, and in such a way that the weight of the cover glass is sustained by them.

SCYPHOZOA and CTENOPHORA.

Members of these classes are peculiarly difficult of preservation and can be studied well only at the sea-shore. The hard parts of coral can be studied in any good collection. The spicules of Alcyonaria (*octocoralla*) are microscopic, and may be isolated as follows :—

38. Place in a test-tube a small portion of the Polyparium of any Alcyonarian. Cover with K-O-H and boil until there is left a residue in the bottom of the tube. Axial portions or other hard masses may be removed and the residue handled by decantation. Allow the residue to settle and then pour off the excess of K-O-II. Fill up with water, shake gently, let it settle and again decant the excess. Repeat the washing with water two or three times and then wash successively in 70%, 95%, 100%, and finally turpentine, leaving the material in the last three for some hours each. Pipette a drop of the material thus prepared upon a slide, drain off the excess of turpentine with a cloth or blotting-paper, add balsam, cover and examine. Different species will yield different results. [This experiment will furnish material sufficient for an entire class.]

Type V.—Turbellaria (any fresh-water form.)

39. *Field-work.* Select for investigation a small pool or ditch of water. The water must be stagnant or slightly flowing, but pure and without foul odors. It should contain an abundance of green water plants and perhaps a little floating green slime as indications of the purity of the water. Use for collecting a net of cheese-cloth and a glass jar. Place the jar in the net and fill it with clear water by reaching out from the margin towards the center of the pool. Let the jar down and pass the empty net slowly back and forth through the water, taking care when turning the net not to reverse it. The minute animals contained in the water will thus be caught in the net and may be transferred to the jar by carefully reversing the net over the mouth of the jar and washing it gently. Hold the jar up to the light and examine. Collect material in the same manner from different depths and portions of the pool and finally scoop up a few of the sticks and leaves from the bottom and add a bit of scum and a piece of some aquatic plant. In this way the jar may be made to contain a sample of the different environments the pool affords. To study the gamma of each environment, the different collections must be kept apart and labelled. In this particular case the material from the bottom will be the most important. Examine the material thus collected and look for soft, very contractile, flat worms, which vary from a short oval to a long worm-like form. They vary in size from almost microscopic dimensions to 1.5^{cm} and in color from slate grey to brown, yellow or green. In an undisturbed aquarium they frequently crawl in an inverted position along the under side of the surface of the water, after the manner of snails.

40. Find a Turbellarian in the material collected in 39, transfer it to a watch crystal with a pipette or glass tube and study with a dissecting microscope. Notice its change of form and its mode of motion. It moves by muscular contraction aided by cilia, which cover its entire surface. These can be well seen on an edge when the specimen is under examination with the high powers [4 c]. In one group (*Dendrocoela*) the intestine is dendritic, i. e. branched like the limbs of a tree, and shows very conspicuously even with the single lens. In others it is a straight rod and not very noticeable (*Rhabdocoela*). Notice also several other organs, usually dendritic, on the sides of the alimentary canal. These are the reproductive and yolk glands. The mouth is on the under side, about central, and rather difficult to see. Two pigment flecks at the anterior end serve as eyes.

41. *General anatomy.* Several common species of fresh water Turbellaria (mostly Rhabdocoela) are so transparent that the internal organs may be well studied in the living animal by compressing it slightly. Cut from a piece of cardboard a frame the size of the cover-glass and 2-3ᵐᵐ wide. Soak it in water and apply it to the center of a slide. Place a living specimen in this with a drop of water, cover and examine. By this means the animal will be slightly flattened and its motions circumscribed. Observe the following—100ᵈ.

a. *Alimentary canal.* The mouth is near the centre of the under side and is best seen by placing an animal ventral side up, although it may be generally made out by focusing through the body. It is circular and may be seen to open and close during the muscular contortions of the animal. The mouth is connected with the main alimentary tract by a cylindrical pharynx, which in a flattened animal may lie unsymmetrically upon the side. The canal itself is rather opaque, owing to its contents, and may be variously shaped [40].

b. *Reproductive system.* Turbellaria are hermaphroditic and the organs of both sexes are very complicated and often dendritic. Two sets of branching organs may be seen at the sides terminating in finger-like lobes. Of these, the thickest lobes, showing white by reflected light, and visible with the simple lens, are the yolk glands. The other set, similar, but more delicate, forms the testes. The uterus, also branched, is very thin walled and almost invisible, but is generally easily located by the large conspicuous eggs which it contains. If these are numerous, they will be seen to be arranged in rows which mark the disposal of the uterine tubules.

c. *Nervous system.* The pigment eyes may be seen to rest upon an opaque mass from which pass four branches, two anteriorly and two posteriorly. The main mass forms the brain, and the branches are the four principal nerve cords. The anterior cords may be followed into the very extensile anterior end, where they resolve themselves into a brush-like mass of minute nerves, which render this extremity a very sensitive tactile organ. The posterior cords run down the sides of the body and divide into minute nerves. There is often a commissure just posterior to the pharynx connecting the two posterior cords.

d. *Nephridial system.* This is an important system, but hard to make out in a living specimen. Four branching tubes, two anterior and two posterior, collect the liquid *excreta* from the body parenchyma. On each side the anterior and posterior tube join and form thus two lateral ducts which run into the pharynx near the mouth.

e. *Cilia.* These are visible only at the edge, but it must be concluded that they cover the entire surface, from the fact that they are visible along the entire edge and at every edge that may be formed by chance foldings and changes of shape. If the animal is sufficiently quiet, a high power (500d) may be used.

42. *Preserving and mounting.* Turbellaria are among the most difficult animals to preserve. It is almost impossible to prevent their complete contraction upon the application of the customary fixing reagents, and even if this be effected, they are apt to disintegrate or dissolve. If the preservation has been successful, there are many obstacles to success in staining. The integument and body parenchyma stain as deeply as the organs, and thus the internal parts, although well stained, are completely hidden. The following methods are an attempt to overcome these obstacles :—

a. *Fixing and preserving.* Use for this *Lang's* fluid, prepared as follows :

Water.	100	parts.
Sodium Chloride,	6-10	"
Acetic Acid,	5-8	"
Corrosive Sublimate,	3-12	"
Alum,	1½	"

Apply it cold, poured suddenly over the animal when expanded. Let it remain in the fluid ½-1 hour, then in 30%, 50%, 2-3 hours each and finally in 70% for preservation.

b. *Staining.* A successfully stained specimen for mounting *in toto* should have the separate systems faintly outlined, the integument and parenchyma being as nearly colorless and transparent as possible. To effect this, two methods may be used—either to stain very slightly, or to stain deeply and afterwards extract the superfluous color with acid. For the first, use ORTH'S Lithium Carmine, Alum Cochineal, or EHRLICH'S Haematoxylin, a few drops in a beaker full of 70%—extremely dilute. Let it remain 1-3 weeks, taking it out from time to time for examination. For the second, use Lithium or Borax Carmine, diluted about one-half, stain 10-15 min, and extract the extra color with acid alcohol (i. e 70% + several drops of 10% HCl.). Watch this and check the action of the acid when necessary, by placing in clear 70%. All specimens for *toto* mounting must be flattened before applying the stain. This may be done with a slide and cover, placing the whole under the lens and pressing with a needle handle. A large specimen, if well hardened, may bear the weight of a slide. Specimens which are to be prepared for sectioning may be diffusely stained without much subsequent withdrawal, and of course should not be flattened. Living specimens, stained in neutral Bismarck brown, as with Infusoria, show clearly defined nephridia and nerve-cords.

43. *Drawing* In this case a diagram is the only practical form of drawing. In all such cases as in that of 40 and 41, where one deals with a rapidly moving object, observation must consist of a long series of rapidly acquired facts and equally rapid conclusions, and the drawing must be a graphic representation of the facts observed. A piece of drawing paper should be at the side of the student, and each conclusion, when clearly formulated, should be sketched in. The details of the diagram should not be all obtained from one procedure : the general shape is best gained by the simple lens : a successful preparation also [42] is of great assistance in comparison with microscopic observation on the living animal.

Type VI.—Distomum (or allied genus).

44. Trematodes, or fluke-worms, are as adults endoparasites in Vertebrates, living in the cavities of the alimentary canal and its accessory organs, the lungs, the gall ducts of the liver, the bladder, etc. They are almost universal in frogs and may be found as follows : Place a live frog under a small bell jar and add a small piece of cotton saturated with chloroform. In a few minutes the frog will be dead and may be removed for dissection. Place it upon its back and with the scissors make a median longitudinal incision on the ventral side, the entire length of the body. The shoulder girdle will be met with in the thoracic region and must be also cut through. Find and examine the following internal organs :

a. *Lungs.* A pair of dark grey sacks lying above the liver on the sides of the oesophagus. If they are empty they will be small, wrinkled and very elastic : if inflated they are large and conspicuous. Cut these off at their base and place in a Stender dish of water for examination.

b. *Urinary bladder.* This lies at the posterior end of the incision, close to the bones of the hip girdle. Look for the large, dark green cloaca, or terminus of the intestine. Upon this may be seen several wrinkled folds of transparent membrane, which spreads over quite a large area, when drawn out with the forceps. This is the bladder. Trematodes may frequently be seen in this, when spread out, owing to the transparency of its walls. Carefully remove it, cutting it away from below, and place it in the dish with the lungs.

c. *Alimentary canal.* This is to be cut through above the stomach and just above the cloaca, and removed entire by cutting away all of its connecting membranes.

These organs may now be searched. Place them one at a time in the small glass dissecting pans on the stage of the dissecting microscope, open them with the scissors, pinning them out when necessary. The entire alimentary canal should be opened and its contents washed out, after which the canal itself may be removed. Each organ has its own separate species of Trematode, each of which have their peculiarities of investigation, and may be separately treated. Any one of these will do for study alive, but the one from the lung may be taken as the type, and is the only one to be mounted.

45. *Study of living Trematodes.* (From frog.)

45. a. *Lung form.* Study this at first superficially in a watch crystal and search for the two sucking discs characteristic of the genus. Then flatten it by compression (it will probably bear treatment between two slides, and perhaps some pressure added to this) and observe by transmitted light, first with the simple lens and then with the low power of the compound microscope. The most conspicuous organ is the uterus, filled with eggs. By a little added pressure the eggs may be made to escape like a fine dust from the genital opening. Locate this and examine eggs with the high power. The anatomy in general is similar to that of Turbellaria [41].

45. b. *The form from the alimentary canal.* This is a very small oval worm scarcely 2ᵐᵐ in length. It is to be placed on a slide in a drop of water, covered by a cover glass and studied with transmitted light. It is very satisfactory to study on account of its small size and transparency. [cf. 41.]

45. c. *The form from the bladder.* This is opaque and cylindrical and should be studied only in a watch crystal of water, attached if possible in its natural attitude to a piece of the bladder-wall. Its ventral sucker is enormous, projecting and cup-shaped, and is used for attachment. The portion of the body anterior to this is very extensile and used as a tactile organ. Compare and homologize the body regions and suckers of this animal with those of the other species.

46. To mount a Trematode use the lung-form, fixing and staining as directed in 42. For rapid work, the use of weak borax-carmine is recommended, followed by the acid alcohol treatment. In this, as in most small Distoma, the testes are not dendritic, but are in the form of compact, rounded masses, which stain deeply, and which, with the similar shaped ovary, form three conspicuous masses posterior to the ventral sucker.

47. *Larval Stages* (=Asexual generations). Trematodes exhibit a complicated alternation of generations, the other forms being parasites of aquatic

animals, generally molluscs, and by a variety of expedients become transferred
to their final Vertebrate host. Collect a few of the common brown snails found
in ponds, break off the shell and examine the soft brownish mass which fills the
apex of the whorl. This mass is the liver and may be frequently found filled
with minute worms of different shapes. These may be transferred to a slide
with a pipette, mounted in a drop of water, covered and examined. The
three main forms are the *sporocyst*, *redia* and *cercaria*, which are very easily
distinguished and may be identified by referring to any text book. (CLAUS or
LANG.) Good permanent mounts are difficult, but may be done by applying
the different reagents drop by drop upon the slide, and removing the liquids
as desired, by inclining the slide and using a cloth or piece of blotting paper.
The entire process should be performed under a dissecting lens. A quantity
of the material may also be prepared at one time by careful decantation
[cf. 38], or perhaps an entire snail's liver may be stained and the separate parasites subsequently dissected out.

Type VII.—Taenia Crassicollis.

48. This is the common Tape-worm of the domestic cat, and is found in
the adult state in the intestines of that animal. To obtain specimens of this,
the intestines of all the cats used in the Vertebrate course are opened with
scissors and examined. They are seldom found in kittens, and are found in
about one out of three adults, there being often 4–6 entire worms in one host.

[If convenient, one or two cats will be examined before the class at this
point, that the Taeniae may be seen in their environment.]

49. *Superficial examination.* Keep the specimen in a glass dissecting
pan, covered with 70% and observe :—

(a) The *Scolex* or " head." This is the blunt point at the anterior end of
the animal, and is not in the shape of a rounded knob with attenuated neck,
as in most tape-worms. [cf. specimens of T. Saginata, etc. Also notice specific
name. *Crassicollis*.] Notice on the Scolex form rounded depressions, the
suckers, also a double row of chitinous hooks. Place the specimen for a mo-
ment upon a slide and bring the Scolex under 50–60ᵈ. Notice the shape and
arrangement of the hooks.

(b) The *Proglottids* or " links." These resemble the segments of segmented
worms, but in reality are not, but are continually produced by vegetative
growth from near the base of the Scolex. The proglottids at the free (posterior) end are the oldest and are sexually mature, while towards the scolex they

grow smaller and less mature, the youngest being at the anterior end where they may be seen dividing off and becoming more distinct. If all the proglottids are broken from the scolex in a living tape-worm, the scolex alone remaining, it will in a few weeks produce a second series of proglottids. On the edge of each proglottid may be seen a slight projection containing a minute opening. This is the genital papilla and is used both for fertilization and for the passage of the eggs. The papillae of some tape-worms occur all on one side, but here they generally alternate, those of adjacent 'links being on opposite sides. In other species each proglottid is double, having genital papillae on each edge.

50. *Preparation of a proglottid.* Select a mature link and cut it off with scissors, leaving portions of adjacent links attached. Place it between two slides, hold it up to the light and apply gradually increasing pressure with the thumbs and forefingers. Watch this procedure carefully and stop at the first indication of cracking. A well preserved link will usually sustain as much pressure as may be applied by this means. As it becomes gradually more transparent, the different organs will become distinct, the most conspicuous being the dendritic *uterus* filled with eggs. A dark line passing from this out to the genital papilla indicates the *oviduct* and the male *vas deferens*. The staining and mounting should follow the rules given in 42.

51. *General anatomy of proglottids.* The anatomical relationships of parts are complicated and difficult to follow. In general the organs are similar to those of Trematodes, altered in shape to fit the rectangular outline of the links, and the lateral position of the genital opening. The successive links represent different stages of development, and hence in order to fully understand the parts, several preparations should be made and compared.

(a) *An "immature" link.* This refers to one taken about in the middle of the animal, where the links are generally much broader than long. In such a link the male organs are mature, while all the female organs, ovary, yolk glands, etc., are small but distinct. Notice: *testes*, many small round masses giving a mottled appearance to the lateral regions of the links; *vas deferens*, a slightly wavy tube leading to the genital papilla and ending in a small pouch containing the *cirrhus* or organ of copulation; *ovary*, an oval mass, central in position, but on the posterior margin; from this last an *oviduct* runs parallel to the *vas deferens*, and opens at the genital papilla; *yolk glands*, lying along the posterior edge of the ovary; *uterus*, a tubular sack extending from the ovary through the center in an anterior direction—this may or may not bear small lateral branches; *nephridia*, a pair of tubes running along the lateral margin and united in each segment by a transverse tube.

(b) *A mature link.* Here the eggs have formed, and filled the uterus, which has branched enormously in order to accommodate them. The other organs, except the oviduct, have become suppressed.

52. *Cross-section of a Proglottid.* This is to be prepared for the class as in the case of the Sycon sponge [22], cut on the microtome and mounted. In this observe: (a) The *body-parenchyma,* filling all the spaces between the organs and leaving no coelom. The parenchyma is crossed by *muscular fibres* which in general run parallel with the body-wall, and surround the organs. Upon the outside of the parenchyma there is a thin transparent *cuticula.* (b) Two lateral *excretory canals* cut across. The section may show a portion of a transverse canal connecting the lateral tubes. (c) The *eggs* which mark the position of the *uterus.* The double contour shows the presence of a thick shell. If the eggs are developed, three pairs of delicate chitinous hooks may may be seen in some, appearing as parallel lines. Such forms are the "six-hooked embryoes," ready to penetrate the walls of the stomach of their first host and encyst themselves in the liver. These hooks are provisional organs and not the same as the definite hooks of the Scolex.

53. *The encysted Scolex* (=*Cysticercus*). This is the Taenia as it develops from the egg, and encysts itself in the body of its *first host.* The cysticercus of this species is found in mice and must be sought for as follows:—Open a mouse by a median ventral incision and either pin it open on the table or bend it gently backwards over the finger. Find the *liver,* a dark red mass consisting of several lobules, lying just beneath the diaphragm a little on the right side. Look over the surface of all the lobes for a large round whitish spot shading off into the red. This spot should be nearly as large as a pencil end and indicates the presence of an oval cyst the size of a pea, imbedded in the liver substance. When such a cyst is found it should be carefully dissected out, and opened under the lens by a small incision at about the middle. Search carefully for a rounded scolex, with hooks and suckers, resembling the scolex of the adult Taenia. When found, carefully remove the superfluous portion of the cyst, preserve the scolex as directed in 42 a, and when thoroughly preserved, flatten slightly and stain with borax carmine as in the case of hydroids [33].

Type VIII.—A Rotifer (especially gen. *Rotifer* or *Brachionus*).

These are microscopic forms found in fresh water, about the size of large ciliate Infusoria and when first discovered, classified with them. They are in reality minute Metazoa with development from eggs. Recently through the examination of a tropical rotifer, *Trochosphaera*, which appears to be a modified Trochozoön, the group has gained in interest, as in it we have probably lineal descendants of that ancient form, differentiated and modified through adaptation to special environments.

54. Rotifers are best studied alive in a drop of water, as in the case of Infusoria. Search in different places in an aquarium, particularly in the slime on the under side of leaves and among the material at the bottom. They are distinguished by an elongated tail, composed generally of joints, and by the characteristic " wheel apparatus," which they unfold at the anterior end, and often suddenly retract. Obtain a good specimen, preferably one whose motion is circumscribed by debris, and observe the following points :—(a) The three body regions : retractile portion or *head*, main body region or *thorax*, and tail or *foot*. (b) The " wheels :" these are in the typical forms a pair of lobes covered with cilia which move in such a way as to give to the lobes the appearance of revolving wheels. These are probably modifications of the prae-oral band or *trochus* of the ancestral form. (c) Notice the stream of particles drawn into the body between the wheels ; follow it inwards and notice the *masticatory apparatus*, a set of chitinous teeth continually moving back and forth. The other organs of the body are difficult to distinguish and may be studied by the help of text books [cf. VOGT u. JUNG. pp. 424-445]. The sexes are separate, but the males are very small and comparatively rare, appearing only at certain times in the year. The females lay two sorts of eggs. The first or *summer eggs* are parthenogenetic, soft shelled, and of two sizes, the larger one producing females and the smaller ones males. The second sort, the *winter eggs*, are the product of copulation ; they have hard shells generally marked with curious and complex sculptures, and produce females alone. In some species (*Brachionus*, etc.) the eggs, when laid, remain clinging to the body of the parent, while in others they are deposited in rows on water plants.

55. *Gastrotricha*. These are a small group of animals, allied to Rotifers and occurring with them. They have a fish-like form, but are depressed and have a broad ciliated double stripe of cilia along the ventral surface, making them appear like hypotrochous Infusoria. They may be considered as modifications of the Trochozoön, but in a different direction from that taken by the Rotifers. In these the trochus has disappeared, the ventral stripe alone persisting as the locomotive organ.

Type IX.—Ascaris lumbricoides.
Ascaris mystax.

The two above species belong to the group of parasitic Nematodes known as "pin-worms" and occurring in the stomach and intestines of Vertebrates. The first named is a large form (20–40ᶜᵐ) occurring in man and the pig, and is generally selected as the type of the group. As this is rather difficult to obtain, it may be supplemented by *A. Mystax*, common in the cat, and very similar to the first, but smaller (4–10ᶜᵐ).

56. *External characteristics (A. Mystax).* For this, use both microscopes according to judgment, the lowest powers first.* Notice the following :—The anterior end is provided with two lateral *wing-like processes*, giving the whole the appearance of the fluke of an anchor. This is characteristic of the species. The *mouth* is situated at the point, surrounded by three *oral papillae.* The posterior end is gently curved in the female, convolute in the male. In both, the *anal orifice* is a transverse slit, just ventral to the tip. It serves also in the male as a reproductive outlet and is furnished with two minute bristles, or *spiculae.* Anterior to the anus upon the ventral side there are 15–20 papillae. In the female the oviducts open by a single orifice, situated in the mid-ventral line at about the anterior third of the body.

57. *Internal Anatomy.* The species *Lumbricoides* is large and easily dissected, but impracticable to use in quantity. *A. Mystax* may be handled in a small dissecting pan, as in the case of the grasshopper, Type XVIII : or a watch crystal may be used, coated with paraffine, and the parts pinned out with fine insect pins. The body should be opened by a median ventral incision, and the body-wall spread apart. The *alimentary canal* is a straight tube extending through the center : it is divisible into an *oesophagus*, somewhat swollen in its posterior portion, and a *chyle-intestine*. The rectal portion receives in the male the vasa deferentia and may be here termed *cloaca*. The *reproductive organs* consist in both sexes of a pair of convoluted tubules, folded back and forth and several times the length of the body. The free end of each is closed, and in this portion the germ-cells are formed. The *nephridia* are contained in two lateral ridges on the body-wall, which run the entire length of the body. They end by transverse ducts near the anterior end,

---

* Hereafter, directions for simple microscopic procedures will not be given, as the student should now be sufficiently skilled in the use of the instrument to be able to exercise independent judgment in each case. An exception to this will be made under Type XVIII, which is to be used as a preliminary study.

which unite and open by a median ventral pore. The nervous system consists of an *oesophageal ring* and a pair of longitudinal cords, dorsal and ventral, which are borne by ridges similar to those which contain the nephridia, but smaller. Both nephridia and nervous cords are best seen in a cross-section [58]. *Circulatory* and *respiratory organs* fail.

58. *Cross-section of Ascaris* (preferably lumbricoides). Look first for *four projections* of the body-wall, 90° apart and corresponding to the four ridges just mentioned. These will serve to orient the section. The *two lateral ridges* are much the largest, and contain each a cross-section of a delicate tube, the nephridium. The *ridges for the nervous cords* support rather than contain them, and thus appear somewhat cup-shaped, bearing a whitish refractive solid substance, the nervous cord. The *body-wall* consists principally of an outer transparent cuticula and a thick mass of irregular muscle-cells, some of which encroach upon the coelom. A fibrous layer between the two and directly underlying the cuticula may be looked upon as a sort of hypodermis, originally cellular, but showing in the adult only a few scattered nuclei. The alimentary canal occupies typically a median position and consists of a single layer of somewhat irregular epithelial cells, with a thin inner cuticula. A muscular layer fails. The remaining tubules, lying in the coelom, belong to the reproductive system and present different aspects in size and cellular formations according to sex, period of development, and regions cut by the plane of the section. In sections of the mature uterus, eggs may frequently be met with, provided with a transparent shell showing a double contour. The embryoes develop to a certain stage in the body of the parent, and may be seen in the early segmentation stages, or in the blastula or gastrula form.

59. *Additional studies of Nematodes.* There are several other forms of considerable general importance, which should be investigated as occasion offers.

a. " Vinegar-eel " (*Anguillula aceti*). Hold a vinegar-cruet up to the light and look for exceedingly minute worms, frequently so numerous as to give the impression of some optical peculiarity in the liquid which causes it to shimmer. These are minute nematodes, feeding upon the fungus which converts the cider into vinegar, and called the "mother." All specimens of vinegar do not contain these nematodes, but if a specimen containing a bit of the fungus be taken, one is almost certain to find them. Pipette a little of such vinegar into a watch crystal and observe with the dissecting microscope. Single nematodes may be transferred to a slide, covered and examined with higher powers.

b. *Trichina spiralis.* This is the famous parasite found in pork, causing in those who eat it a dangerous muscular disease. It was the cause assigned by Germany for prohibiting during several years the importation of American pork. The parasite exists in infected meat in the encysted form, the separate cysts appearing as minute white dots, somewhat oval in form, occurring between the separate fibres. Specimens of infected meat may often be obtained from Governmental experiment stations or from private investigators. Tease out a few scraps of the meat under the lens and look for the cysts. They are better seen by reflected light. To study them they are best flattened gently, and stained with borax carmine [42 b]. They should remain in the turpentine until quite transparent, when the worms will be seen coiled up in the cysts.

c. *Nematode parasites in the frog, etc.* Interesting parasitic forms may be found in the lungs and intestines of the frogs, and generally appear while searching for Trematodes [44]. The black lung-form is especially interesting, being an hermaphroditic parasitic form, which alternates with a bi-sexual form living in damp earth. This phenomenon is known as *heterogony.* (cf. the text-books under *Rhabdonema nigrovenosum* and *Rhabditis nigrovenosa.*) A little nematode found in the frog's intestine is very transparent and convenient for microscopic study.

Type X.—Lumbricus Terrestris.

60. *Study of a living specimen.* Place a large earth-worm in a saucer of water, and notice the following :—It elongates and contracts in length. What sort of muscles cause this? What is the course of their fibres? It changes its caliber at the same time. What sort of muscles do this? What difference in shape of the two ends? Which is anterior? Which end is the more sensitive? Where must the nervous system be best developed? At which end are the sense-organs best developed in all bilateral animals? Why? Has the animal distinct dorsal, ventral and lateral aspects? Notice along the sides, rows of minute bristles, especially prominent posteriorly. If the animal be drawn between the fingers, there is more resistance in one direction than another. Can that be explained by the direction of the bristles? Notice the *cingulum,* a fleshy girdle surrounding the body. Is this nearer the anterior or the posterior end? Notice the body-rings or *segments.* These are true segments and not reduplicated abdomens as in the tape-worm.

61. *Killing and preserving.*

a. *Killing.* While dying, earth-worms are apt to contract and throw themselves into such tight coils that dissection is impossible. To obviate this, place a worm in a saucer of water and introduce alcohol very slowly until the desired result be obtained. It answers nearly as well and is often more convenient to throw them alive into 25–30%. They will writhe and coil themselves at first, but ultimately the majority will straighten out quite well.

b. *Preserving.* If the specimens are to be used for dissection, the body cavity should be opened by a few slits along the mid-dorsal line, and the specimen returned to the 30% 4–6 hours and then in 70% for permanent preservation. If immediate dissection is desired, open the animal as soon as killed, dissect under 30% and place in 70% at the end of the first day's work. For staining and sectioning, a more complicated course is necessary. The alimentary canal is normally filled with earth, which renders sectioning impossible. To get rid of this, keep a living animal for several days in a box of damp coffee-grounds until the earth is entirely replaced. Fine saw-dust is also used, but is not as good. Fair results may often be obtained by simply starving an earth-worm in clear water, changing it daily. In sections taken from such animals, the cells of the alimentary canal will be abnormally contracted. If none of the above methods are convenient, very small earth-worms may be used, thus reducing the quantity of earth to a minimum. When ready, the animal is to be killed as above, but removed as soon as possible, and cut into several pieces, which are to be separately preserved in some good fixative with subsequent treatment with alcohol. Either of the following are especially recommended :—

I. Aqueous corrosive sublimate 2–6 hours.
    Water : running, or large quantity often changed, $\frac{1}{4}$–$1\frac{1}{2}$ h.
    30% alcohol 4–8 h.
    70% alcohol—permanently.

II. Kleinenberg's Picro-sulphuric acid. 2–6 h.
    70%—This should be changed repeatedly at intervals of 4–12 h., until it remains colorless.

The original location of each piece should be known, and may be designated by cutting the pieces into different lengths, or by preserving them in separate bottles, carefully labelled.

62. *General Anatomy.*

(a) *External characteristics.* Notice again on the dead animal the parts mentioned in 60, the segments, bristles, cingulum, etc. At the anterior end is a partial segment, projecting over the mouth. This is the *prostomium*, or lip, and is not counted as a segment. Count the segments from this past the

7

clitellum; what segments does the clitellum include? Posterior to this the segments vary in number. Notice upon the ventral side, anterior to the clitellum, a pair of slit-like openings, surrounded by fleshy lips. These are the external openings of the *vasa deferentia*. Upon what segment are these? Look for a pair of minute openings upon the segment just anterior to this. These are the openings of the *oviducts*, and are visible only during the period of oviposition. Between the 9th and 10th, and between the 10th and 11th segments, are two pairs of minute openings leading into *spermathecae*, or receptacles for the spermatic fluid. These are visible only during the pairing season. Each segment possesses a pair of *nephridia*, which open on the ventral side, at the anterior border of each segment, in front of the little pits for the bristles. These are at all times difficult to see. The *anus* is at the posterior extremity of the body.

(b) *Internal anatomy*. The dissection should be under alcohol in a glass dissecting pan, as in the case of Type XVIII. The animal may be cut in two about 10–15 segments posterior to the clitellum, and only the anterior portion studied, as the posterior segments are all alike. An incision should be made with the scissors along the mid-dorsal line, and the animal carefully pinned out. The segments are separated by connective tissue partitions, the *dissepiments*, which are attached to the integument. These prevent one from spreading of the body-walls properly, and should be cut from the integument with a scalpel. When this is done, make a superficial examination of the organs *in situ*, without farther dissection. Notice :—The *alimentary canal* with its divisions, running through the center.—The *dorsal blood vessel*, lying upon it, filled with red blood—a series of 5-6 pairs of *enlarged lateral blood vessels*, the so-called " hearts," also red, which are given off from the dorsal vessel.—Between these, 2–3 pairs of little orange bodies, the *calciferous* or *oesophageal* glands.—The *spermatic vesicle*, a large mass of whitish lobes, three on each side.—The *spermothecae*, two pairs of little round sacs, just anterior to the testes and lying on the ventral wall.—The brain, or *supra-oesophageal ganglion*, a dumb-bell shaped whitish mass at the extreme anterior end, and lying upon the anterior enlargement of the alimentary canal.—The *nephridia*, little coiled tubes lying on the sides of the body, in each segment. These are best seen in the posterior portion. Notice farther that the dissepiments are complete partitions, dividing the segments into separate coeloms, and allowing continuous organs, as the alimentary canal, to perforate them, as in the diaphragm of mammals.

After this general survey, the systems are to be carefully dissected, one at a time, cutting merely the connective tissue connections, studying the shape and location of each part, and the relations of parts to each other. In this, reference may be made to the following tabulated view of the systems. For details, refer to the different text books.

63. *Anatomical Synopsis.*

[In this, the order is that of preparation : the numbers refer to the segments.]

1. ALIMENTARY CANAL.

Mouth. With prostomium.

Pharynx. A large oval mass, attached by muscular fibres to the body-wall.

Oesophagus. A long, narrow tube, surrounded by the hearts (5-13), and bearing the—

Oesophageal, or calciferous glands. These are three pairs of lateral diverticula of the oesophagus, containing calcareous bodies (11-13).

Crop (Ingluvies). An oval shaped dilatation of the posterior end of the oesophagus. Its walls are comparatively thin (14-16).

Gizzard, or stomach (Ventriculus). A round mass with thick, muscular walls (17-18).

Intestine. A large tube of equal caliber throughout, with thin walls disposed laterally in folds. A portion of the dorsal wall is invaginated, forming the *Typhlosole.* (see cross-section) 19—to end.

Liver mass (chloragogue cells). A diffused mass of yellow-brown pyriform cells, covering the intestine dorsally and giving it its yellow color. (Seen in cross-section : also by isolating the cells by scraping the canal very gently, and mounting temporarily.)

2. REPRODUCTIVE SYSTEM (Hermaphroditic).

(a) Male.

Vesicula Seminalis (spermatic vesicle). A large rectangular, yellow-white organ, lying beneath and at the sides of the alimentary canal (about 10-11), varying in extent in different stages of development. Divided across into two portions : the anterior expanded laterally into two paired lobes, and the posterior into one. These lobes increase in size posteriorly, and when well developed are recurved upon the main portion.

Testes. These are beneath the spermatic vesicle and are exposed by removing the dorsal wall of the vesicle and taking out its contents. They may then be seen as two pairs of little solid white organs, lying upon the sides of the nerve cord (10, 11). Their cells pass by some unknown way into the spermatic vesicle, in which they develop.

Vasa deferentia. Slender tubes, beginning with two pairs of funnel-shaped openings just posterior to the testes (10, 11). Upon each side the two tubes unite into one (12), thus forming two lateral vasa deferentia, which open externally (15). The funnels collect the ripe spermatozoa which escape from the spermatic vesicle by an orifice communicating with the coelom.

(b) Female.

Spermothecae. Two pairs of spherical sacs, used to contain the spermatic fluid of *another individual*, received during pairing. The eggs are fertilized by this fluid and not by that of the same animal. They lie upon the ventral wall (9, 10) and open separately between the segments (9–10) (10–11).

Ovaries. A pair of very small organs, lying upon the ventral wall (13) on each side of the nerve cord. The mature eggs break loose and float about in the coelomic cavity until collected by the oviducts. The ovaries are scarcely visible except when mature.

Oviducts. Two little tubes, which begin as funnel-shaped openings in the dissepiment between (13) and (14). These collect the eggs and open externally in (14).

3. CIRCULATORY SYSTEM.

The blood is red (colored plasma, white corpuscles) and thus the vessels are easily traced. It is a closed system, entirely separate from the coelom.

Longitudinal vessels. These are five in number. (1) median dorsal, (2) median ventral, (3) sub-neural, (4) and (5) a pair of lateral neural vessels. The blood flows in an anterior direction in (1), posteriorily in the others.

Commissural vessels. The lateral stems communicate by vessels which pass around the body. In the genital region, five pairs of pulsating commissures, the "hearts," pass from the dorsal to the ventral vessel. In the intestinal region, commissures unite the dorsal with the sub-neural. Other communications are through anastomoses of capillary terminations.

Lateral supply vessels. The body-wall, intestines and other organs are supplied by vessels passing from the main trunks. These form capillary net-works about the parts supplied.

4. EXCRETORY SYSTEM.

Every segment except the first three possesses a pair of *lateral Nephridia* (segmental organs). They are the largest in the segments just posterior to the reproductive organs, and are best studied there. Each nephridium is a convoluted tube, having a free end, which floats in the coelomic cavity. This has a *ciliated funnel-shaped opening* and collects excretory material from the coelom. The tube passes through the dissepiment and empties externally in the succeeding segment. Each tube may be divided into a *transparent, a glandular,* and a *terminal* portion.

5. NERVOUS SYSTEM.

This consists of a *brain* or *supra-oesophageal ganglion,* lying upon the pharynx, and a *ventral cord* possessing *ganglionic enlargements* in each segment. The brain sends out *anterior nerves* to the prostomium and unites with the ventral chain by a pair of *lateral commissures.* These send out nerves to the pharyngeal region. From the cord pass out lateral nerves, a *ganglionic pair* from each ganglion, and an *interganglionic pair* from the intervals between them.

64. *Study of a cross-section.* Mount a microtome section cut from a piece prepared as in 61, and try to interpret the parts shown by means of the anatomical knowledge gained by the dissection. The parts especially brought out will be the typhlosole, the longitudinal blood-vessels, the chloragogue cells, and the nervous cord. The study of the body-walls and the muscular layers is only to be accomplished by sections. If the plane of section passes through the bristles (*setae*), examine and compare with 65.

65. *Study of parapodia of a Polychaete (Nereis).* This is the common clam-worm found at low tide in mud-flats. Notice in an entire specimen the lateral expansions of each segment which bear the setae. These are the *par-*

apodia. Cut off a single parapodium with the scissors and examine in a watch crystal. It may afterwards be flattened and mounted. The parapodium is double, consisting of dorsal and ventral portions. On the dorsal half notice :—A flat triangular piece, the *gill*—connected with this a slender process. the *cirrus*—a fleshy lobe bearing several bristles (*setae*), and a single median spine (*aciculum*). Compare the ventral half. What part is lacking?

Type XI.—Cambarus sp ?*

(*Astacus* or other cray-fish will do as well. *Homarus* may be used for class demonstrations.)

66. *External anatomy.* Cray-fish may be killed by placing them under a bell-jar and adding a bit of cotton soaked with chloroform ; or by immersion for a few moments in strong alcohol. In the latter case they should be removed as soon as dead, to prevent the shrinking of the tissues which would occur if the alcohol were allowed to penetrate the interior. The specimens may be examined in the air, but should be dropped into water at least once an hour, that the internal parts may remain moist. A jar containing one or two living specimens should be placed on each table for study of the use of external parts. The parts may be studied in the following order :

I. GENERAL FORM.

> *Bilateral symmetry.* with dorsal, ventral, two lateral, anterior and posterior aspects. What other types have been bilateral ? Is the distinction between dorsal and ventral more or less in *Lumbricus ?* in *Ascaris ?* Which shows higher development ?

> *Skeleton.* This is an *exo-skeleton.* consisting of a chitinous cuticula, reinforced by mineral salts. Test the carapace with a drop of HCl. What does it prove ? Notice the structure of the joints (*arthra*), delicate membranes extend between the hard pieces and allow free motion. Notice how the hard pieces overlap when parts are contracted. cf. dorsal pieces of abdomen. Compare the whole with mediaeval armor.

---

* The arrangement of Crustacean types is the reverse of the usual order and is rather that of convenience in obtaining material, than that of logical sequence. It is supposed that Type XI will be studied at the opening of the college year (Sept.–Oct.), and is followed by forms that may be kept alive in laboratory aquaria, and by marine forms that must be studied in alcohol. The natural order, showing the derivation from polychaetous Annelids, would be the following :—(1) *Polychaete.* (2) *Branchipus.* (3) *Cyclops.* (4) *Porcellio.* (5) *Cambarus.*

Metamerism. Possesses true segments (somites or metameres). Are the segments similar or dissimilar? What is the advantage of metameric differentiation? Is it greater or less here than in *Lumbricus?* Which is the higher?

Body-regions (=*segment-complexes*). Three body-regions as in grasshopper : first two fused, making (1) a *cephalo-thorax*, composed of immovable segments, and covered dorsally by a single shield-like piece, the *dorsal carapace*, and (2) an *abdomen* of moveable segments.

Cephalo-thorax. A transverse suture (*cervical suture*) near the middle. This marks division between head and thorax. A median, anterior process, the *rostrum*. The sides of the carapace are free and form covering pieces for the gills (*branchiostegites*). Break off one and notice the gills underneath : also the lateral body-wall, delicate and white, behind them. The gill chamber is thus external. Ventrally, notice the appendages, five pairs of legs, of which the first is for attack and defense, and not used in walking. The genital openings are at the basal joints of a certain pair of legs, differing in the two sexes. Compare different specimens. Where are they in the male? in the female? Find the mouth and note the large number of appendages about it, used partly as jaws and partly to hold or taste the food.

Abdomen. Consists of seven segments, each with a large dorsal plate, *tergite*, and a very narrow ventral piece, the *sternite*. The first five segments bear small legs used for swimming (*pleopods*). The appendages of the sixth segment are expanded, and form, with the seventh segment, which is without appendages, a five lobed tail. How is this tail used? In the spring the females attach the eggs and young to the pleopods and protect them with the tail. The seventh or final segment bears the *anus.*

II. Appendages.

There are nineteen pairs of appendages, each of which represents a somite. Of these pairs, thirteen belong to the thorax and six to the abdomen. The cephalo-thorax thus consists of thirteen somites and the abdomen of seven (the last without appendages), giving a total of twenty somites. They are best studied in the following order. Remove them from one side only, leaving those of the other in their natural order for comparison.

1. *A typical pleopod.* Select this from the 3d–5th abdominal somite. It consists of a basal piece, *protopodite*, double in this case : and of two terminal branches, the outer or *exopodite*, and inner or *endopodite*. This is considered a typical appendage, of which the others may be considered modifications.

2. *The sixth pleopod.* This is flattened to assist in the formation of the tail. One of the terminal branches is double. Which is it?

3. *The first and second pleopods.* These are different in the two sexes. In the male they are large and peculiarly twisted, and capable of approximation to form a copulatory tube for the conveyance of the spermatozoa. In the female they are both normal in shape, but the first are very small.

4. *A walking leg (periopod).* Select the next to the last leg. Separate it from the body, taking care to remove it entire. It consists of seven joints, to the first of which are attached a plumose gill and a flat gill paddle. *epipodite* or *flabellum.* The joints are named in order : (1) Coxopodite. (2) Basipodite. (3) Ischiopodite. (4) Meropodite. (5) Carpopodite. (6) Propodite. (7) Dactylopodite. Select the first leg beyond the large claw. Notice that it ends in a claw also i. e. is *chelate (chela=*claw). Does this result from the addition of an *eighth* piece? Compare with the former one and explain. Compare with the large claw. Is the difference in size the only one? cf. living animal for difference in use. Compare with the typical form (a pleopod). The leg may be explained by the suppression of one of the terminal branches and subdivision of the other into joints. To find out which branch is suppressed, compare with the next.

5. *The third maxilliped.* This is the segment next anterior to the large chelae. (The segments will now be taken in order, going towards the anterior end.) This is intermediate between a jaw and a foot, as its name denotes. Has it an epipodite? A gill? Do you recognize the leg part? Has it the same number of joints as the others? Do you find the branch which was not developed in the true legs? How would you then describe a périopod?

6. *The second and first maxillipeds.* Remove these in order and compare with the third. Which terminal branch shows a gradual reduction? Which one shows increase? Can you homologize the parts of the endopodite in each? Look for a gill and an epodite in each. Notice in the first, the flattened maxillary processes, projecting inwards. There are two of these, corresponding to the two divisions of the protopodite.

7. *The second maxilla.* Do you recognize in this an exopodite? An endopodite? An epipodite? With what is the epipodite fused? How many maxillary processes are there? Can you explain them with reference to those of the preceding?

8. *The first maxilla.* Which of the two terminal branches fails here? Can you homologize the maxillary processes?

9. *The mandible.* This consists of a hard jaw with toothed edges and a palpus. What typical part does each represent? Compare with last.

10. *The second antennae.* These consist of a protopodite and the two terminal branches. Which one is developed into the long, tactile portion? Notice at the base, the conspicuous opening of the nephridial organ, the so-called "green gland," situated in the center of a raised papilla.

11. *The first antennae (antennulae).* These are biramous. Upon one of the branches are situated tufts of club-shaped hairs, probably olfactory. Examine both branches with the microscope and find out which one possesses them. Examine the base of these antennae for the auditory organs, little cavities open to the air, into which the animal pushes sand-grains to serve as otoliths.

12. *The eyes.* These have been supposed to have the value of appendages, being movably articulated with the body, but they develop as fixed parts and become secondarily free.

The number and relationships of appendages and somites in Crustacea may be expressed by formulae. An $\times$ denotes that a part is present ; R signifies that a part is rudimentary ; — that it fails. Simple abbreviations may be used for the names of the appendages. Roman numerals for the somites of the cephalo-thorax. Arabic for the abdomen. The following is the formula for Cambarus:—

No. of Somite.	Name of Appendage.	Basal Piece.	Exopodite.	Endopodite.
I	Ant_1	×	×	×
II	Ant_2	×	R	×
III	Md	×	—	×
IV	Mx_1	×	—	×
V	Mx_2	×	×	×
VI	Mxp_1	×	×	×
VII	Mxp_2	×	×	×
VIII	Mxp_3	×	×	×
IX	Leg_1(Chelae)	×	—	×
X	Leg_2	×	—	×
XI	Leg_3 { gen. op. / female. }	×	—	×
XII	Leg_4	×	—	×
XIII	Leg_5 { gen. op. / male. }	×	—	×
14 15	Pl_1 } modified Pl_2 } in male.	×	×	×
16	Pl_3	×	×	×
17	Pl_4	×	×	×
18	Pl_5	×	×	×
19	Pl_6 (caudal fin)	×	×	×
20	—	—	—	—

III. GILLS.

Remove the branchiostegite from the side upon which the appendages have not been disturbed, and examine the delicate plumose gills. For this, immersion in water or weak alcohol will be necessary. There are three classes of gills, named according to their point of attachment.

(1) *Podobranchiae*, attached to the basal joints of the legs.

(2) *Arthrobranchiae*, attached to the delicate membrane between the skeletal pieces : of these there are two sorts, anterior and posterior.

(3) *Pleurobranchiae*, attached to the lateral body-wall (*Pleuron*).

Gills may occur on every segment between VI-XIII, and each segment may have four, one podobranchia, two arthrobranchiae (anterior and posterior) and one pleurobranchia. Thus typically a complete equipment would number $4 \times 8 = 32$. Actually, however, a portion of these fail, the resulting arrangement being constant in a given species, but varying in different species. Fill in the following blanks with the formula for *Cambarus*, indicating the presence of a well-developed gill by an ✕, a rudiment by R, and adding the epipodites.

GILL FORMULA FOR CAMBARUS.

Somite.	Podo.	Ant. Arth.	Post. Arth.	Pleuro.	Epip.	Total.
VI	——	——	——	——	——	——
VII	——	——	——	——	——	——
VIII	——	——	——	——	——	——
IX	——	——	——	——	——	——
X	——	——	——	——	——	——
XI	——	——	——	——	——	——
XII	——	——	——	——	——	——
XIII	——	——	——	——	——	——
Total	——	——	——	——	——	——

67. *Dissection of internal organs.* Cambarus is dissected from the dorsal side as in other Arthropods. For an animal of this size, a small tin dissecting pan should be used, and the specimen covered with weak alcohol. A lens may be removed from the dissecting microscope and held in the fingers over the object, when minute observation is required. If necessary, a small and complicated part may be removed to a small glass dissecting pan, or watch crystal, and examined microscopically.

The specimen should first be taken in the hand and the dorsal carapace carefully removed in small pieces, taking great care not to harm the subjacent organs. Then the abdominal tergites may be cut through at the sides and removed. When this is completed, pin the specimen in the pan.

[The student should now be ready to identify the different organs himself, at least in a simple form like Cambarus, and hence general rules for identification of parts and order of dissection will be substituted for the usual anatomical description. This method will be followed as far as possible in the rest of the book, excepting, of course, Type XVIII, which is designed as a preliminary study.]

As a general rule, the *alimentary canal* should first be sought, and whenever it is coiled or in any way obscures other parts, may be dissected away from its connections and laid aside. In this case, however, the delicate central organ of the *circulatory system*, the "heart," lies above the intestine, and should first be studied. The circulation can only be successfully demonstrated in an artificially injected specimen, and for this purpose a large form, like the lobster, should be used. With the alimentary canal occur various digestive glands, one of which, the *liver*, is very constant in its occurrence. It is to be expected below the stomach, and is generally brownish or greenish in color and voluminous in size. Its connection with the alimentary canal should be traced, after which it may be removed with a portion of the canal. The *stomach* is peculiar in this animal and possesses a set of chitinous teeth, worked by external muscles. This should be carefully examined. In a bisexual form, as in this case, the sex should be determined by external indications if possible, and the internal *reproductive organs* should then be sought. As essential parts of this, one may expect a germ gland, *ovary* or *testis*, and a tube, generally long, and often coiled, through which the germ cells may pass to the exterior. In the male this tube is proportionally small, and is termed *vas deferens*, while in the female it is often large, and termed *oviduct*. If a portion of it is enlarged to serve as a receptacle for the retention of eggs or embryoes, it is called the *uterus*. Accessory organs are common and may be always expected, in the male, glands to secrete a liquid medium in which to suspend the germ cells, and in the female, glands for yolk, shell-material, albumen, etc., besides receptacles for spermatozoa, etc.

Nephridia, either as separate tubes or united into a mass, may be expected in any location, and in this case are in the form of the green gland, at the base of the second Antennae. The *nervous system* is similar to those of the earth-worm and grasshopper. The *muscular system* should be studied in connection with the skeleton, whether external or internal. The mechanism of a few joints, or of a few characteristic body-movements, will prove sufficient in the general study of a type.

68. *Histological study.* To obtain and preserve material for the study of cell structure, an animal should be killed by chloroform, and dissected either in the air or under water. The parts to be sectioned should be selected, cut out as rapidly as possible, and preserved by the method recommended for pieces of earth-worms, 61 b, I and II. The pieces selected should be small, and separated as much as possible to allow the ready access of the reagents to all parts. After preservation, the parts may be kept in small bottles of 70 % well labelled. For sectioning, a piece may be stained *in toto*, the principal being the same as in the staining of hydroids, the amount of time being regulated by the size and density of the piece. From the turpentine the piece is put into melted paraffine (in the paraffine oven) and imbedded when completely infiltrated. It is impossible to assign exact times for the different procedures.

but the following may be taken as a mean estimate. Specimens are not harmed by exceeding the time in the stain or in 70%, but the higher alcohols and the turpentine render them hard and brittle.

Preparing an object in toto for microtome sections.

1. A specimen, preserved as above, is taken from the bottle of 70% and placed in borax carmine—twenty-four hours.

2. Take from the stain and place in " acid 70% " i. e. 70% + a few drops of HCl. Watch the piece and see it turn from a dull maroon to a bright scarlet. The change should be moderately slow (5-10 minutes). A too rapid change denotes an excess of acid, a too gradual change denotes too little. This should be remedied in either case. The piece should remain until it is thoroughly penetrated by the acid alcohol (10-20 minutes).

3. Place in clean 70% (2-6 hours).

4. 95% alcohol (6-12 hours).

5. 100% alcohol (4-8 hours).

6. Turpentine (4-8 hours).

7. Melted paraffine (2-4 hours).

8. Imbed.

Small Stender dishes are the best to use for the work of the fluids. The 100% should be kept in a tightly corked bottle. Bottles are often conveniently used for carrying about when changes are to be made out of laboratory hours. A preparation made in this way is amply sufficient for an entire class, and may be made by one person, the students taking turn in thus preparing the class specimens. Sections through the "liver" (*better hepato-pancreas*) or through the intestine are very instructive and simple in structure.

Type XII.—Porcellio sp?

69. *Collecting and preserving.* This is the common "sow-bug" or "damp bug," a terrestrial Crustacean found in damp shady woods beneath stones, and is often met with under bricks and boards in barns and cellars. Simple immersion in 70% is sufficient to kill and preserve them for the study of the external parts, and each student before the approach of cold weather should collect a bottle full of such specimens (8–10) to serve as a supply. For internal anatomy they should be dissected in a fresh state, also prepared for the microtome and sectioned in various planes. As with insects, their hard chitinous exo-skeleton forms a barrier to the penetration of the various preservatives, to overcome which, various means have been adopted, as follows :—

1. The use of *hot* fixing reagents. These are to be heated in a test tube and poured suddenly over the animal while alive. Perhaps the best for this purpose is hot alcoholic corrosive sublimate. This should remain 20–40 min., then washed out in 70 %.

2. The selection of specimens which have just moulted and in which the chitin is soft and tender. These are also best preserved with hot reagents.

3. Punch or prick a hole through the exo-skeleton in a region not afterwards to be used for sectioning. A living animal may be cut in two with a sharp razor and the two parts instantly dropped into the fixative.

As in the case of preservation, staining also presents serious difficulties. To overcome these, use an alcoholic stain of a high grade (Borax Carmine may be made as high as 70 %); also continue the immersion for a long time (several days or weeks if necessary). This will not hurt the specimen if the stain contains a high percentage of alcohol. Sections may be cut from an unstained object and stained on the slide, by the method known as "slide staining." For this, cover the slide with an *albumen fixative*, rubbing it in well with the finger. After this, heat in the usual way, until the paraffine is melted, and then apply the usual reagents *in the reverse order* (4–5 minutes each), turpentine—100 %—95 %—70 %, etc., until the specimen reaches the grade of the desired stain. Apply the stain to the object and then pass it up again through the same succession as far as the turpentine. Then wipe the slide carefully, except the area covered by the specimen, which may be well drained but *must not become dry*. Add a drop of Balsam and cover as usual. The application of the different reagents is best performed by immersing the entire slide successively in jars containing the desired reagents, or in the case of single sections, the reagents may be pipetted on the slide, drop by drop, and successively wiped away, held in the hand or rested from time to time on some convenient support. All the different stains may be applied in this way, haematoxylin being especially recommended.

70. *External Anatomy.* Use for this specimens killed by simple immersion in strong alcohol. Remove them, and spread them out on a glass slide upon the stage of a dissecting microscope, allowing them to dry. The larger external parts are best studied in this way, while the smaller appendages, maxillae, etc., may be mounted temporarily in a drop of water for details. Look over several specimens and distinguish the following : (1) Females carrying young. These have a brood-sack formed by broad plates attached to the inner side of the first five pairs of legs. In this may be found eggs and young in all stages. Such females possess a median birth-opening between the fifth and sixth pairs of legs, while the lateral genital openings are obsolete. (2) Females without brood-sack. These lack the appendages upon the legs, also the median birth-opening, but possess lateral genital openings at the bases of the fifth pair of legs. Females change from one of the above forms to the other by a moult. (3) Males. These possess a median organ of copulation at the juncture of thorax and abdomen. This folds backwards with the gills, but is generally easily distinguished by its darker color. A pair of genital openings lie at its base.

1. *General body-form.* The body is *depressed,* i. e. flattened derso-ventrally. The thoracic somites are free, and not united into a carapace as in Type XI. On the dorsal side note :— *The head :* this consists of the true head of five somites + the first thoracic somite : the *long antennae* (=second): look carefully on the inner side of the base of these for the very rudimentary *first pair :* the *eyes,* a facetted surface on each side. *Seven free thoracic somites :* the first is concave anteriorly and receives the head, the last concave posteriorly, for the small abdomen. *The abdomen,* six separate pieces, large lateral processes on the second—fourth, the sixth representing the sixth and seventh somites fused. cf. Type XI.

2. *The appendages.* These may be separated with forceps or sharp needle, the larger ones studied dry and the smaller ones mounted temporarily in water. Those of the thorax and abdomen should be studied first, upon the entire specimen. For the parts of the head, remove this piece by cutting it through from the dorsal side. The parts here are delicate and may be best handled by placing the entire piece in a watch crystal and covering it with water. Make a detailed study of the separate parts, as in the case of the previous Type, and draw each. Refer to the following formula :

Somites.		Appendages.	
I		Ant$_1$	Rudimentary.
II		Ant$_2$	Large.
III	Fused.	Md	
IV		Mx$_1$	
V		Mx$_2$	
VI		Mxp	Lamellate, folded over the others.
VII		L$_1$	
VIII		L$_2$	
IX		L$_3$	With brood lamellae in female.
X		L$_4$	
XI		L$_5$	
XII		L$_6$	
XIII		L$_7$	
14		Pl$_1$	
15		Pl$_2$	
16		Pl$_3$	Endopodite serves as gill.
17		Pl$_4$	Expodite large, containing air chambers. Serves also as protection for the true gill.
18		Pl$_5$	
19	Fused	Pl$_6$	Without gill, forming part of tail.
20		———	Central tail piece, fused with 19.

Head. — somites I–VI (Fused)
Thorax. — somites VII–XIII
Abdomen. — somites 14–20

71. *Study of a cross-section.* The internal anatomy of *Porcellio* is very similar to that of Type XI, and need not receive especial attention. A general idea of the relationship of the different parts may be obtained from a cross-section prepared as in 69. The section should be taken through the thorax and will serve as a type to illustrate the relationship of systems in Arthropods in general. In the center will be the alimentary canal ; above this will be the dorsal blood vessel, and beneath it the nervous cord. The tubules and glands of the reproductive system will lie symmetrically upon the sides. If the section be well cut, the relation between the muscular masses and the exo-skeletal plates will be beautifully shown. The external chitin is difficult to cut and is apt to crack into little pieces, some of which will be carried upon the edge of the knife across the section. This may be obviated by the use of the following reagent :

> Gum mastic,
> Absolute alcohol.
> Ether,
> Collodion.

Take about equal parts of these, put the gum mastic in the alcohol, and allow it to dissolve as much as it will : then add the other ingredients.

This is to be applied to the flat section-surface of the paraffine block, and allowed to dry. The microtome screw is then turned and the section cut off in the usual way. The liquid forms a thin film over the section which holds the pieces of chitin in their natural place.

72. *Preliminary study of the next three types.* These represent three groups of minute Crustacea, having an average size of a small pin-head, and constituting the bulk of the material collected with the tow net, whether in marine or fresh water. The type specimens used here are fresh water forms and are collected in the same way as the Turbellaria. Type V. They often develop in quantity in the laboratory aquaria through the chance introduction of a few adult individuals, or from eggs brought in with the mud. Place the material containing these in a shallow crystallizing dish and set over a black surface. Look for small, rapidly moving animals, which may be distinguished by shape and mode of motion.

1. Somewhat elongated forms, body tapering behind into a tail. The females often possess a pair of egg-sacs half as large as the body, depending from the sides of the abdomen. They move by quick, darting motions, often progressing 3-4 inches by a single propulsion. They are thus very difficult of capture and are best taken by a long glass tube, into which a quantity of water may be suddenly drawn. These animals are Copepods, of which the commonest form is *Cyclops*, Type XIII.

2. Oval forms, somewhat flattened laterally, i. e. compressed. (cf. description of *Porcellio*, 70, 1.). The body is covered by a pair of lateral shells which leave the head free. They are propelled by the second pair of antennae, which are enormously developed and used as

oars. Hence their motion consists of short rapid jerks. There are many common species of this group; Branchiopoda, Sub-order *Cladocera*, any one of which may be taken as a type. The form described below is *Simocephalus*, a common form, larger than the average, and should be secured if possible.

3. Perfectly oval forms, entirely covered by large lateral shells, which include the head. These are the smallest of all and seek the bottom and sides of the glass, progressing by a steady, rolling motion, caused by the projection of minute legs from between the edges of the shells. They are Ostracods, mostly belonging to the genus *Cypris*, which is taken as Type XV. The large species, variegated with green and white, should be taken, if possible.

The methods of investigation are similar in all three types. The two first are best studied alive, as they are so nearly transparent that the internal organs are readily seen. Examination in a watch-crystal under the dissecting lens will give the general body form, the use of the appendages, mode of motion, etc., while they may be placed under the compound microscope by using slide and cover. They may be quieted by chloroform, which may be applied in drops, or by holding near them a bit of blotting paper, saturated with it. Warm Perenyi's fluid poured over a group of *Cypris*, placed in a watch-crystal will cause them to die with the shells beautifully expanded. They may be preserved and sectioned as in 69. A mass of them may be handled at once by filtering the water containing them through a bit of muslin, after which they should be tied up in the muslin by a thread, and the bag and all subjected to the action of the different reagents. They must be removed from the bag to be placed in the paraffine oven and may be imbedded by pouring the paraffine containing them into a watch-crystal previously smeared with glycerine. The position of the separate individuals may be seen through the translucent mass and individuals that lie in a favorable position may be cut out in a cubic or oblong piece, which may then be melted upon the end of a large block and cut in the usual way. They are rather difficult to stain *in toto*. Slide staining gives good results.

Type XIII.—Cyclops sp?

73. *External anatomy.* The body is covered dorsally by a carapace which includes the head (five somites) and one leg-bearing thoracic somite. Then follow four free thoracic somites, each bearing a pair of legs. The long tapering abdomen consists of five somites (first two fused in female) and ends in a fork (*furca*) tipped with long plumose bristles. In the center of the anterior end of the carapace is the eye, a fleck of pigment which gives sensation of light and possibly of color. This eye gave the animal its name, but it is really double, as is seen by its shape. There are two pairs of antennae, of which the first bears tufts of olfactory hairs. It is transformed in the male into a clasping organ, used during pairing. The legs consist of a double basal piece and two branches, each consisting of flattened joints. There are no gills. The formula for somites and appendages is as follows.

Body Regions.	Somites.	Appendages.	
	I	Ant$_1$	(Clasping in male.)
	II	Ant$_2$	
Head	III	Md	
	IV	Mx$_1$	
	V	Mx$_2$	
	VI	L$_1$	
	VII	L$_2$	
Thorax	VIII	L$_3$	
	IX	L$_4$	
	X	L$_5$	(Reduced in size.)
	11	———	
	12	———	
Abdomen	13	———	
	14	———	
	15	———	(With anus and furca.)

74. *Internal anatomy.* The *alimentary canal* is a straight tube in the median line with but little differentiation of parts. Anus in fifth abdominal segment. No accessory organs. During life the canal oscillates continually back and forth, stirring the fluid in the body cavity and thus functionally replacing a *circulatory system*, which fails entirely. There is no special respiratory system. The *reproductive* organs differ in appearance during different periods and should be studied in connection with the life history and development.

75. *Development.* This should be studied by arranging a small aquarium in a Stender dish or tumbler. Place in the bottom a layer of pond mud and introduce a bit of some green alga. Cyclops will multiply in this readily and very rapidly, and all stages may be found. For special study, individuals, pairs, etc., must be isolated and kept in a smaller dish. Investigate the following points :—

(a) *Males.* These are much smaller than the females and of rather rare occurrence, only appearing at certain times. They are recognized by the peculiar modification of the first antennae. Testes and vas deferens are recognizable only at time of pairing. The reproductive openings are upon the first abdominal somite. The spermatozoa form masses known as *spermatophores,* a pair of which appear during pairing, as little oval refractive bodies depending from the genital somite.

(b) *Females.* Look over several females until you find one in which the ovaries appear as dark branching masses. These lie symmetrically on either side of the alimentary canal, connected by one or two commissures. After fertilization, the eggs pass out from these through openings in the first abdominal somite, and form a pair of external egg-masses, the *ovisacs,* attached by stalks to the parent.

(c) *Pairing.* This may be observed only through chance, by continued watching during the period in which males are abundant. The male clasps the first antennae about the fifth pair of legs of the female, and directs a strong current of water towards her by a rapid rowing motion of the legs. The two spermatophores, which appear on the outside, become liberated by this and are carried across to the female, to which they become attached. The ovisacs form soon after this (15 min.-1 hour), taking but a few minutes for their formation.

(d) *Larvae.* The young free-swimming larvae of Cyclops are called *Naupliae,* and are of theoretical importance from the occurrence of similar larvae among widely different groups of Crustacea, thus furnishing support for the theory that the Nauplias represents the primitive Crustacean form. They are common in aquaria filled with females, and are often seen in the field of the microscope during the study of the adults. A Nauplius is oval in form and possesses three pairs of appendages used as legs. In later development these become the first and second antennae and mandibles, while new somites and appendages appear posterior to the original three, and develop into the remainder of the body.

[76. *Parasitic Copepoda.* Many members of this group have forsaken their independent existence and are found as parasites upon the gills or external surface of aquatic vertebrates, representing all stages of degeneracy. Open the *operculum* or gill-flap of any fish, marine or fresh water, and search among the reddish fringes (gills) for these forms. Some are irregular, almost shapeless masses, tubular or sac-like, while others still show more or less affinity to Cyclops. Even the most degenerate possess the lateral egg-sacs which, however, may appear as close coils of tubules. When found, cut away the portion to which they are attached, wash until clean and examine with lens. Ascertain mode of attachment, looking especially for hooks, burrs, or other parts used in clinging to their host. Are these modifications of typical parts, or new formations?]

Type XIV.—Simocephalus sp ? (or other typical Branchiopod).

77. *General anatomy.* The general structure may be ascertained from living specimens placed under a cover-glass. They lie upon their side and are flat and transparent, giving one the opportunity of using the higher lens. For mode of motion, shape as seen from above, etc., they must be studied in a watch-crystal.

(a) *External anatomy.* The *shell* is formed by an integumental duplicature, that is, it is an exaggeration of a lateral fold. Focus on its surface for peculiar markings upon it. The *head* is free and provided with a short beak, the *rostrum.* The *abdomen* is slender and without appendages. It moves rapidly back and forth, being capable of projecting from beyond the margin of the shell. It is tipped with a pair of curved terminal spines, above which is a row of smaller spines. It is generally held in a curved position along the ventral side. The *eye* is median (its true position being seen in a dorsal view) and consists of a mass of black pigment, surrounded by highly refractive bodies, the "crystalline lenses." The eye is moved by minute muscular bands. There are two pairs of *antennae*, a short anterior pair, near the rostrum, provided with knobbed *olfactory hairs*, and an enormously developed second pair, the *oar-antennae*. These possess two branches, one of four and the other of three joints. They are used for locomotion and thus functionally replace the *legs* which bear *gills* and are used solely for respiration. These legs are seen through the shell, waving back and forth, thus beating the gills through the water. There are five pairs of legs, of which the last is placed at some little distance behind the others. The mouth parts consist of a pair of mandibles and two pairs of maxillae, but are too minute for general study.

(b) *Internal anatomy.* The *alimentary canal* is somewhat curved, terminating in the anal orifice. Where situated? Two small "livers" or *hepatic diverticula* lie in the head. Look at these from above. Notice the color of the food in different parts of the canal. Examine a crushed specimen with high powers and ascertain of what the food consists. In the neck region are a few very delicate wavy tubes, running transverse to the longitudinal axis of the body. These form the "shell gland," an excretory organ, consisting of a mass of nephridia, which open near the second maxillae. The *nervous system* consists of a brain lying back of the eye, in the head, connected by commissures to a chain of seven ventral ganglia. Dorsally, at the beginning of the thorax, lies a little oval pulsating organ, *the heart.* It possesses a pair of lateral ostia for the reception of the blood, which is thence forced outward anteriorly. There are no blood-vessels. The reproductive system is closely connected with organs for the care of the young and the development, and should form the subject of separate investigation.

78. Specimens may be kept in small aquaria, as in the case of Cyclops, 75. Notice the following :—

(a) *Males.* These appear only at certain times in the year, the females reproducing at other times parthenogenetically. The *testes* are simple tubes ventral and parallel to the alimentary canal, tapering posteriorly into *vasa deferentia,* which open ventrally in the abdominal region. The males are best distinguished by the absence of the brood cavity of the female.

(b) *Females.* These are the common form, being often the only sex present in a large aquarium. The *ovaries* lie in a similar position to the testes. They are nearly transparent, but may be distinguished by large cells, the eggs, which characterize them. The *oviducts* are at the posterior end of these tubes, but open dorsally into a broad cavity, and not ventrally, as in the case of the vasa deferentia. The *brood cavity* is an oblong space, bounded above and at the sides by the dorsal shell, ventrally by the alimentary canal, and is closed posteriorly by a pair of processes, which project dorsally from the abdomen, and which may be drawn downward to allow the escape of the young. The brood cavity is filled with a special fluid for the nourishment of the young, which develop here without metamorphosis and escape in the same form as the adult.

(c) *Eggs* and *ephippium.* The eggs which develop parthenogenetically are thin shelled, develop very rapidly and produce only females. Such eggs are called "summer" eggs. The fertilized or "winter" eggs are hard-shelled and receive, when in the brood cavity, a second covering, called the *saddle* or *ephippium,* which is produced by the shell. An ephippium contains in this species two eggs, and is deposited in the mud, in which state the eggs can survive the winter, or endure drying up. Winter eggs produce both sexes.

Type XV.—Cypris sp?

79. This is one of the most refractory objects of study, being too small to dissect, too opaque to treat like the former two types, and covered with thick chitinous shells, rendering it difficult to cut. It is best in a general course to be content with a superficial examination. They may be first studied alive, observing the movements of the shells, and of the appendages. For special study of the latter, a large specimen may be selected and one lateral shell lifted off with needles, or they may be opened by treatment with hot Perenyi's fluid [72]. The shell is closed by a transverse adductor muscle, which unites the two valves at their centers. When this muscle is relaxed, an elastic ligament, situated at the dorsal hinge, draws the shells apart. An impression for the attachment of this muscle may be seen on the inner surface of a clean, empty shell. There are only seven pairs of appendages, borne on the anterior body portion, which corresponds to head and thorax; the abdomen is attenuated and resembles the appendages. The appendages, with their forms and uses, are as follows :

Ant_1) Long, leg-like, used in locomotion. The second pair modified
Ant_2)   in male.
Md   Large—provided with a palpus, and flattened piece or fan-plate.
Mx_1 With a palpus and flattened fan-plate.
Mx_2 With a small, two-jointed palpus, and rudimentary fan-plate.
Leg_1 This is often called the *maxilliped.* Used for locomotion.
Leg_2 " Cleaning foot," curved backward and used in cleaning out the
        shell.

80. *Development.* Cypris develops by a Nauplias, which is a fine illustration of caenogenetic modification, or that modification by which an important modern character develops during the repetition of an ancestral stage. Here the Nauplias is provided with the characteristic bivalve shell. Search at the bottom of an aquarium filled with Cypris, and examine the smallest forms. The shells are transparent and some of them may be seen to possess but three pairs of appendages, the number characteristic of Naupliae.

81. *Comparison with barnacle larvae.* The common barnacles, which cover the rocks and all submerged objects at the sea-coasts, are modified Crustacea, similar to Ostracods, which develop as free-swimming Naupliae, and afterwards become sessile in an inverted position. When sessile, the shells develop calcareous plates, and the appendages grow into delicate tendril-like extremities, suitable to direct to the mouth a current of sea water, containing nutritious material. The Nauplias larva is similar to that of the Ostracods, and a later stage, the so-called "Cypris" stage, is similar to the adult. Barnacles may thus be considered modified Ostracods. The young stages are commonly found in marine tow, which should be diluted and examined as in the case of Medusoids. The Naupliae of marine Copepods, similar to Cyclops, are also to be expected, but these are distinguished by the absence of lateral shells.

Type XVI.—A Spider (almost any form will do).

82. *External characteristics.* Spiders may be killed by chloroform, benzine, or by immersion in alcohol. All spiders are venomous, but only the large ones have mandibles powerful enough to penetrate the thick cuticle at the finger tips. They can bite only what is beneath them, and may thus be firmly grasped from above with the thumb and finger. They may often be induced to drop into a wide mouthed bottle filled with alcohol, by holding it beneath them, and poking them carefully with a stick. For external parts, alcoholic specimens may be used, but large fresh specimens must be selected for dissection. Investigate the following external characteristics :—

   (a) *Body.* Two divisions, a cephalo-thorax and an abdomen. Anterior end square with abrupt anterior edge, upon which are situated simple lenticular eyes, eight in most spiders. The arrangement and relative size of these is an important systematic distinction. Compare and draw several different forms. The waist is attenuated and the abdomen is generally so convex dorsally that it laps over the poste-

rior border of the thorax. Neither piece shows segmentation,
although the paired appendages indicate that it is a segmented animal. Is this consolidation a primitive characteristic, or is it a modification? Is the spider a high or low form?

(b) *Appendages.* The cephalo-thorax possesses six pairs of appendages,
of which one is prae-oral.

    1. *Chelicers (mandibles).* These depend downwards from the
abrupt anterior end. They consist of a powerful basal joint
and a movable tooth or poison fang. A poison gland in the
interior connects by a fine duct with this fang. In some
members of this class the chelicers bear a double claw.

    2. *Pedipalpi* (palpi). These resemble legs in the female, but are
modified in the male into spoon-shaped organs, used to convey the spermatozoa over to the female during pairing. In
some forms these also may become huge double claws (Ex.
Scorpion).

    3-6. True legs. These are seven jointed and possess minute
toothed claws at their extremity. These are used in running
about on the web.

The abdomen of the embryo possesses rudimentary appendages,
some of which disappear, while others form the spinnerets or spin-
ning glands, at the end of the abdomen. How many pairs are
there? In cases where a median spinneret is present, how may it
be explained?

(c) *Respiratory and genital openings.* Spiders breathe by a system of
plates, or *fan-tracheae*, which hang in respiratory chambers, the so-
called "*lungs.*" A few tropical forms have two pairs of these, but
North American spiders have only a single pair. Look at the ventral side of the abdomen, close to the cephalo-thorax, for a pair of
smooth, often shiny plates. These are the *opercula,* which cover the
respiratory chambers. This system is reinforced by a system of simple tracheal tubes, which open by a single median *stigma* just anterior to the spinnerets. The female possesses a median opening as
outlet for the eggs, and two lateral openings for the reception of the
spermatozoa, and leading into spermothecae. The median opening,
or *vagina,* lies between or just posterior to the anterior pair of spinnerets, and the lateral openings at its sides. The *vasa deferentia* of
the male open by a median opening between the opercula.

83. *Anatomical Synopsis.*

[A spider should be dissected in the same general way as other Arthropods ; the specimen opened from the dorsal side by cutting around the eyes and removing the entire dorsal wall, and then dissected under weak alcohol, pinned down in a small glass pan. As the specimens should be large and fresh, it is impossible to provide material for a large class. Students finding good specimens may make a general dissection, employing the following synopsis :]

1. *Alimentary canal.* Muscular pharynx—long, narrow oesophagus, running through a mass of nerve ganglia—thoracic enlargement with paired diverticula reaching into legs and running around to the ventral side—an abdominal portion, also showing a few diverticula, (This, together with the thoracic enlargement, forms the chyle-stomach.)—the short rectum and large rectal vesicle, a sort of cloaca—a pair of Malpighian or urinary tubules which empty into the rectum.

2. *Nervous system.* An almost solid mass, perforated by the oesophagus and giving out paired nerves to the eyes, appendages, abdominal viscera, etc. The portion dorsal to the oesophagus represents a supra-oesophageal ganglion and the ventral portion, the consolidated ventral chain. Some forms have a single small ventral ganglion just posterior to the main mass.

3. *Circulatory system.* A dorsal vessel or "heart" in the median line of the abdomen, with three pairs of ostia to receive the blood from the body, and anterior, posterior and lateral aortae, to send it to the different parts. These arteries are short and with open ends, the remainder of the circulation being lacunar.

4. *Respiratory system.* Fan tracheae consisting of hanging plates with double walls. The blood passes down between the walls of each plate and is aerated by the supply of air in the respiratory chambers. Tracheal tubes somewhat similar to those of insects, a single median tube arising from the single stigma, which branches into two main stems, furnished with little tufts of tracheal tubes.

5. *Reproductive system.* *Male :* Two long tubular testes in abdomen, becoming gradually more attenuated to form the vasa deferentia, which are often much convoluted. They unite at the end and open by a median orifice. The pedipalpi serve as copulatory organs. The inner side of the terminal joint contains a spiral tube opening at the apex. This is filled with spermatozoa, which are thus conveyed to the receptacula of the female. *Female :* The two ovaries are tubu-

lar, but filled with large masses of egg-follicles, giving the whole the appearance of a bunch of grapes. The short oviducts are continuous with these and unite with each other near the median opening to form a short vagina. Generally two receptacula seminis situated in close proximity to the median opening. In both sexes the blind ends of the tubular germ glands may unite, forming a sort of ring.

84. *Development.* Egg coccoons, formed by the mother, may be found during the winter upon trees, and in summer among the webs. They are of different species, but may be used indifferently. Examine for winter coccoons any little rolled leaf or bit of leaf attached to the twigs, and look among bushes for large round hanging masses. The detailed study of the development requires careful technical manipulation, but a general idea may be gained by studying the eggs with a lens or with the compound microscope. During the early stages the eggs are opaque, but in later development the growth of the embryo may be seen, the budding appendages, the appearance of the eyes, etc.

Type **XVII**. A **Chilopod** (preferably *Lithobius*).

85. *Collecting and preserving.* Myriapods are the "thousand legged worms," found under the bark of decayed stumps, under logs, etc., in damp woods, and occasionally in damp soil. The two orders are distinguished by the legs, one having a single and one a double pair for each somite. Those referred to here (*Chilopoda*) have a single pair for each somite and are somewhat depressed in form. Alcoholic specimens are sufficient for the work required here, which is merely the external anatomy. The internal anatomy is very similar to that of insects, and the mode of investigation the same.

86. *External anatomy.* Note the well marked segmentation. Is there distinction between body regions? Are the somites differentiated? Is this higher or lower than a grasshopper? Study the parts in the following order :

1. *The head.* This may be removed from the body and studied in a watch crystal. Be careful not to include the large curved pair of jaws which are reflected upwards and lie over the mouth. They are accessory mouth-parts, but belong to the thorax. The separate appendages may be temporarily mounted for special examination. The eyes are small masses of simple eyes (ocelli) on the sides of the head. One pair of antennae. Mouth parts consist of mandi-

bles, and first and second maxillae. (Remove these in the reverse order for convenience and study separately.) Compare each maxilla with the typical first maxilla of the grasshopper, consisting of basal piece, inner and outer plate, and palpus. In what way does the second maxilla differ from the type? The first maxilla?

2. *The body* (=*thorax* + *abdomen*). Study the appendages of the first body somite, often called the maxilliped. Notice the inner plate with its teeth, and the free leg. What is the probable use of this free part? Notice a small opening on the inner side near the point. This is the opening of a poison gland. Compare this maxilliped with the second maxilla and with a free leg. To what does the palpus correspond? Study a typical leg. How many joints? Can you name them after the homology of the insect leg? [Type XVIII.] Look at the last four legs for coxal glands. Is the last pair of legs any different? Why? Compare with the anal stylets of crickets and cockroaches. The somite posterior to the one bearing the last pair of legs is the genital somite, in which is the median genital opening. The appendages are changed into accessory genital organs. Are there differences in the two sexes? The final or anal somite is without appendages and contains the anal orifice.

Type XVIII.—Caloptenus, sp?

[Dissosteira (*Oedipoda*) the large form, with black and yellow wings, or any other true grasshopper (*Acrididae*) may be used here.]

[As this is intended to be used as a preliminary study, the rules for dissection and other technique will be given in full. Anatomical details, as relation, size and appearance of parts, will also be given.]

87. *Collecting, choice of material, preservation, etc.* The majority of grasshoppers are hatched in the early summer and grow continuously until Sept.-Oct., when swarms of large adults may be found everywhere. Obtain a few large forms, which may be caught in the fingers or with a small net, and select from them those having the following characteristics : *Body brownish above, yellow beneath ; shanks of hind legs bright red : a conical spine projecting from the throat, just beneath the chin.* This last is the most important characteristic and indicates the group of grasshoppers known as *Caloptenus.* If this form cannot be obtained, some other large species may be used, but one must expect to find some slight variation from the description given here.

The sexes may be distinguished as follows : *Male*, smaller than female, with darker colored body, terminating posteriorly in a large rounded knob. *Female*, often twice the size of the male, with large heavy body terminating in four spines, which may approximate to form a single sharp spine for boring holes in the ground to deposit the eggs. The sexes may be indifferently used as specimens, except in the sections on the reproductive system. For immediate use, or for the study of external parts, specimens may be dropped alive into collecting bottles containing 70–80 % alcohol. This will not penetrate the interior, however, sufficiently to preserve the inner organs. Benzine is the quickest and simplest killing agent and may be used whenever a specimen is to be dissected at once. Dried specimens are often useful in the study of external parts.

88. *Symmetry. Form. Orientation :* The body is *bilateral*, i. e. it may be divided by one and only one plane, into two symmetrical halves. Locate this plane. The halves are termed *right* and *left*. Imagine the animal resting on a level surface in the natural position, and pass a plane through it, parallel to the level surface and perpendicular to the first plane. Are these halves equal and symmetrical? What are the characteristics of each, i. e. color, general shape, thickness of shell, etc.? What external environments may have produced the difference?

The upper half is termed *dorsal*, the lower, *ventral*. Pass a plane through the middle of the body, perpendicular to the two preceding. This divides the body into two halves, *anterior* and *posterior*. The above terms are also applied relatively, thus the inner pair of wings are ventral to the outer, although both lie in the dorsal half. The exterior is hard, formed of plates which overlap each other, allowing the necessary motion, as in a suit of mediaeval armor. This hard exterior is really the skeleton, and is termed *exo-skeleton* in distinction from an internal, or *endo-skeleton*. The joints overlap each other in the direction from which injuries are liable to come, or in the direction of the most usual motion. cf. the joints of the posterior half of the body.

89. *Metamerism. Body regions.* The body is divided transversely into more or less separate joints or *segments*. The segments are also termed *metameres* or *somites*, and an animal possessing them is said to be *metameric* or *segmented*. cf. oyster and earth-worm. In a simple metameric animal, the segments are all alike and distinct from each other, while in higher forms segments may be suppressed, over developed, united together and modified in many ways. This condition is considered higher because it has developed from the first, and is for the purpose of doing certain things better ; thus the middle ventral region of the body may be consolidated into a single plate to offer a better support for the legs. Is the grasshopper a higher or lower metameric animal? In higher forms the segments may be grouped into three body regions or segment complexes, the *head*, *thorax* and *abdomen*, each equipped with its peculiar parts, and fitted to perform a certain range of functions. The head bears the sense organs, antennae, eyes, etc., and possesses internally the central organ of the nervous system, the brain. Its function is therefore *psychic*. The thorax bears the legs and wings, and contains the largest of the body-muscles for moving these parts. It thus controls the *animal* functions. The abdomen bears the organs for egg-laying, fertilizing the eggs, etc., and contains the reproductive and digestive organs. These functions are shared by plants and are termed *vegetative*.

90. *Morphology. Its methods.* Morphology is the comparative study of anatomical details. It considers every animal a plastic form, starting from a simple primitive type, and moulded and modified to suit its peculiar environments and conditions of life. It thus seeks to express every anatomical detail in terms of the typical form, rather than by means which are purely arbitrary. This may be made plain by referring to the following postulates, applicable to the group of animals under consideration.

1. *Typical somite.*[*] The grasshopper is too much modified to obtain from it a clear idea of a simple unmodified somite ; but the study of more primitive forms yields the following conception :—An approximately circular ring, composed mainly of two skeletal pieces, a larger, overlapping dorsal piece, the *tergite*, and a smaller, ventral piece, the *sternite*. These are united by lateral membranes, in which a farther pair of pieces may develop, the *pleurites* or *pleura*. This somite bears a pair of *jointed appendages*, inserted between the pleurite and sternite.

2. *Typical animal* (of this group). Corresponding to the above definition, a typical animal would be formed by a *series of such somites*, each overlapping the next posterior. Each somite must be of the same size and shape and must contain the same organs. There are no actual animals corresponding exactly to this description, but the more primitive members of this group approach nearer to it. cf. *Chilopoda* and *Peripatus*.

3. *Modifications.* The grasshopper may be considered as the result of a series of gradually changing environments brought to bear upon the type

---

[*] The above description is that of skeletal parts merely.

form, and must be compared with it, part by part. The most common modifications of the type are the following :—

(a.) *Excessive development of a part.* Ex. The tergite of the first segment of the body. In some species this extends beyond the end of the abdomen.

(b.) *Reduction of a part.* Ex. The sternite of the same segment. The reduction may lead to complete loss.

(c.) *Change of form to subserve a certain function.* Ex. The antennae, jaws and legs are all appendages.

(d.) *Fusion of several parts.* This is done wherever support is needed, either for some powerful muscle, or against extensive injury. Ex. The partial fusion of the sternites in the thorax, or the complete fusion of the head to protect the brain. This change often affects the somites themselves so that their identification is often difficult.

91. *Rules for resolution of fused somites.*

1. As each pair of appendages represents a somite, their number will correspond to the *minimum* number of somites. Why the minimum? This is called the "Rule of Savigny."

2. Often internal parts retain metameric structure longer than external. Why? Which parts are more directly under the influence of external conditions? The nervous system is especially primitive in this regard.

3. Individual development records past conditions more or less perfectly, and thus parts fused in the adult may be separate in the embryo or the larval form.

92. *Special study of external details.* Place a specimen upon the stage of a dissecting microscope and examine with the lowest lens, controlling the work with a pair of forceps in the left hand and a stout needle in the right. For minute details use the higher lens. For some purposes the lens is best removed from the microscope and used like a hand lens. If a specimen has just been removed from a bottle of alcohol it should become dry before the investigation to avoid the confusing refraction from the liquid.

1. *Head.* Remove this from the body, place it face towards you upon the stage and notice :—The *antennae*, delicate organs of touch. How many joints? The two large *eyes* at the sides. These are "compound," composed of numerous facettes. Look at these with the

higher power. The *ocelli* or simple eyes, one median, upon the ridge between the antennae, and two lateral, at the inner, upper edge of the compound eyes. The *labrum*, or upper lip, covering the mouth. Lift this up and notice the two heavy black jaws, the *mandibles*. These move laterally, and are toothed on the opposing inner surfaces. At the sides of the mandibles are four jointed tactile organs, the *palpi*, attached to accessory mouth parts, not visible from above. Look over the general surface for seams or *sutures* dividing the head into separate areas. Of these the lateral pieces are the *genae* or cheeks, the small median one just above the labrum is the *clypeus*, and the remainder the *epicranium*. Make a large outline drawing of the head from the front, putting in these details.

2. *Mouth parts.* Turn the head over and separate the loose pieces about the mouth. The first (lowest) is a double piece with two palpi, then will follow two lateral pieces, each resembling a half of the first piece, and each with one palpus, and lastly the mandibles. These pieces should be separately studied by placing them in a watch crystal of 70 ℒ alcohol, and using the dissecting lens. The separate lateral pieces are the *maxillae*, and the double piece the *labium* or *second maxillae*. Between them all is found a little fleshy projection, the tongue. These parts, together with the labrum, constitute the " mouth parts," the arrangement of which is as follows :

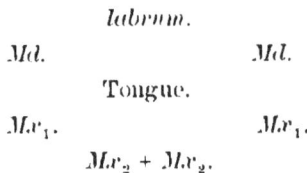

labrum.

Md.　　　　　　　　*Md.*

Tongue.

Mx₁.　　　　　　　*Mx₁.*

Mx₂ + *Mx₂.*

All of these, with the exception of the tongue, are believed to represent appendages. The labrum is plainly a pair united in the middle, and the notch in the center of the free margin of the labrum denotes its double origin. Make a detailed drawing of the first maxilla and the labium, noticing that each consists of a basal piece (two jointed) two terminal plates, inner and outer, and a palpus. The palpi are termed respectively, *maxillary palpi* and *labial palpi.*

3. *Segmentation of the head.* As the fusion is complete, this is done entirely by counting the appendages, proving these, in doubtful cases,

by their relationship to pairs of nerves which supply them. The present solution is as follows:

Somite.	Appendages.
I	Antennae.
II	Lateral ocelli.
III	Median ocellus (=a fused pair).
IV	Eyes.
V	Labrum.
VI	Mandibles.
VII	First Maxillae.
VIII	Second Maxillae (labium).

4. *Thorax.* This consists of three segments, termed respectively, *pro-*, *meso-* and *meta-thorax*. Their tergites are termed *pro-*, *meso-* and *meta-notum*, and their sternites, *pro-*, *meso-* and *meta-sternum*. The terms ending in -ite, used in defining the parts of the type, are thus only employed in the abdomen. The pro-thorax is free and movable. It consists of a broad pro-notum, forming a sort of collar, and a very narrow pro-sternum, bearing a spine, the *pro-thoracic tubercle*. Its appendages are the first pair of legs. The meso- and meta-sternum are fused to give a firm support to the third or jumping legs. They are reinforced by a portion of the first abdominal somite. Dorsally they are divided up into a complex mass of small skeletal elements, to give freedom of motion to the two pairs of wings which are not appendages, but integumental folds which have become movably articulated with the rest of the skeleton. Ventrally a sternal plate is formed, the sutures of which show its composition. The most anterior piece is the *meso-sternum*, into which the *meta-sternum* is fitted by a joint known as a "dove-tail." The *first abdominal sternite* is again dove-tailed into it, and forms the posterior portion of the sternal plate. Make an outline drawing of the sternal plate. The *meso-* and *meta-thoracic legs* are inserted at the posterior edges of their respective segments. The *pleura* of the meso- and meta-thorax are double, an *episternum* in contact with the sternal plate, and an *epimeron* above it. There are also a few minute pieces developed about the bases of the legs.

5. *Legs.* The first four are used for walking and the last pair for leaping. They consist of the same parts, but their relative size differs according to function. The two long joints are the "thigh" (*femur*)

9

and "shank" (*tibia*), the angle between them being termed the
"knee." Between the body and the thigh are interposed two little
joints, *coxa* and *trochanter*. Below the tibia extends the foot or
tarsus. This typically consists of five joints, the terminal one bearing a pair of claws (*ungues*), but in the grasshopper the number is
reduced to three. Examine the *pulvilli*, or little pads, at the bottom
of the feet and see if you can explain this. Draw one of the legs.

6. *Abdomen.* The somites here conform nearly to the type, but lack
appendages and pleurites. They consist of *tergite* and *sternite*, and
are regular and plain except the first, which is fused ventrally with
the thorax, and at the end, where they are modified to accommodate
the reproductive organs and the outlet of the intestine. In this they
differ in the two sexes. In the female are four long hooks, the *rhabdites* or egg-guides. The ducts of the germ glands, and the alimentary canal have their outlets among these terminal plates. Keep this
in mind and find the relationship of these parts, during the dissection
of these systems. In the dorsal portion of the first segment, underneath the wing, will be seen the *ear*, a vibratory membrane stretched
across a cavity.

93. *Digestive system.* For all internal dissections, use fresh specimens
and as large as possible. Hold the animal in the left hand with the head
toward you, and clip off with the scissors the legs, wings and antennae.
Then insert the point of the scissors just beneath the anterior edge of the pro-
notum, at the place where it bends over laterally, and cut through the exo-
skeleton, keeping on the side, as far as the end of the abdomen. Make a similar incision on the other side. In this, keep the concealed point of the instrument as near as possible the inner surface of the exo-skeleton, in order not to
wound the parts beneath. Then pin the animal to the wax of a small glass
dissecting pan, running the pins through the incision, close to the lateral
walls. Cover the specimen with 30 % and place the pan upon a dissecting
microscope. Take the forceps in the left hand and the scalpel in the right,
and lift off the free dorsal piece, cutting off the connections as close as possible to the part to be removed. The internal organs will be seen lying in
place in the ventral portion of the exo-skeleton, which thus serves as a shell.
The alcohol renders transparent parts opaque, and the use of a liquid in dissection is for the purpose of unfolding delicate membranes, etc., which would
remain in air as a confused heap. (cf. the use of water in mounting "sea
mosses.") If the pins be gently spread out, the walls will widen a little. The

dissection consists of separating the different parts by cutting and teasing the almost invisible connecting bands, and is best done by the forceps and the sharp needle (the little spear-pointed instrument). The lens may be turned on or off according to desire, and the microscope stage makes a good rest for the little dissecting pan. Look first for the *alimentary canal*, a greenish or brownish tube running through the median axis of the body. It is divided into the following parts : *Buccal* or *mouth cavity*. This lies in the head and is not seen in this dissection. It is merely the little cavity within and between the mandibles. It is easily distinguished by its very dark brown walls. *Oesophagus*, a narrow tube running straight upwards to the apex of the head. There it makes a turn toward the neck. After passing the neck, the canal suddenly swells out into a huge *crop*, tapering at both ends and filled generally with a dark brown fluid, which consists of the food mixed with some of the digestive juices. This is the "molasses" which is disgorged to inspire fear in its enemies. At the base of the crop the canal is re-inforced by six finger-like expansions, which possess both anterior and posterior portions. These are the *gastric diverticula*. They are hollow and thus reinforce the general digestive surface. They empty into a portion of the canal which forms the boundary between the crop and the stomach. It is called the *proventriculus*, and is developed in the cricket and others, into a little spherical bag containing six hard, chitinous teeth. The *stomach* or *ventriculus*, is that portion of the canal which extends from the diverticula to the place where the canal is joined by countless little tubes appearing in a sort of tangle. They are the *urinary* or *Malpighian tubules*, which absorb excretory material from the body cavity in which they lie, and pour it into the lower portion of the canal. How many separate bunches are there? The remainder of the canal is termed the *intestine*, and may be divided into three portions, the *ileum*, or tapering conical portion, the *colon*, of small size, and slightly coiled, and the *rectum*, with six rectal glands.

The plate above the anus is termed the anal plate. Which is it? Draw the canal and its parts, putting in the outlines of the body. Remove the entire alimentary canal, cutting it through at its two ends, and place in a watch crystal. Examine then the floor of the thorax for whitish masses something in the form of bunches of grapes. These are the salivary glands, which form anteriorly two tubes or ducts, opening into the mouth.

Return now to the alimentary canal : cut it open with the scissors, rinse out the contents and examine the inner walls. The crop is marked by ridges, which are surmounted by tiny teeth, pointing backward, to prevent regurgitation of the solid food. (cf. later directions for mounting this.) Does the proventriculus give any indication of teeth as seen in the cricket? Can you find the inner openings for the gastric diverticula? For the urinary tubules? What else do you find at the beginning of the intestine?

94. *Female reproductive system.* Select for this a large female specimen [87] and prepare as for digestive system. The large yellow masses lying dorsal to the intestine are the *ovarial tubules*, or organs for the formation of the eggs. Part these in the middle and float them out at the sides of the body, without cutting the tube with which they are connected. The mass will resolve itself under the lens into a collection of tapering tubules inserted by their larger end into a large tube, the *oviduct*. How many separate ovarial tubules compose one of the lateral masses? What is the condition of the upper, free end of the oviduct? Trace the two oviducts down until you find them uniting into one. What relation does this median tube sustain to the alimentary canal? Trace it out to the external orifice and locate it. What relation does it sustain to the rhabdites? The lower portion of the alimentary canal may be removed if in the way. Look near the union of the two oviducts, and find a little pyriform sac, connected by a long duct with the exterior. How is this connection made? The sac is the *Receptaculum seminis,* which becomes filled during pairing with male germ-cells (*spermatozoa*), which may then fertilize the separate eggs as they come down into the median tube. Draw the entire system, putting in enough of the other parts to show relationship. Cut off a separate tubule, remove the large egg at the end, and place in the center of a clean slide. Put on a drop of the liquid used in dissection, and lay on a cover-glass. If there are air bubbles, add more of the liquid, drop by drop, until the cover-glass is full. If there is too much liquid, the cover-glass will float and the excess of liquid must be withdrawn by applying a cloth or bit of blotting paper. Such a preparation is termed a *temporary mount.* Examine this with the compound microscope 60–80ᵈ. [As this instrument is first introduced in this place, its use must be explained.] Of what does the tubule consist? What idea do you gain concerning the formation of eggs? Draw the specimen. [Stained preparations and sections will give further information on this subject and will be studied later.]

95. *Male reproductive system.* This consists, like the female system, of a system of tubules which secrete the germ cells, connected with excurrent ducts. Prepare a male specimen as above, and examine the orange yellow mass lying in the abdomen above the intestine. This is the *testis,* and may be resolved into a double bundle of nearly transparent club-shaped sacs, surrounded by a lace-work of yellow fat. The two bundles of *spermatic tubules* may be separated as in the case of the ovary, and found to connect with a small tubule, the *vas deferens*.

The two vasa deferentia unite below into a common tube, as in the case of the oviducts. A mass of coiled white tubules lying upon either side, connect also with the common tube. How may this mass be resolved? The tubules composing it may be termed *prostate* tubules and secrete a fluid medium in which the minute germ-cells may be suspended. Make a drawing illustrating this system. The finer anatomy of the tubules must be followed by sections, as will be directed later.

96. *Respiratory system.* In the insect the air is brought into intimate relation to all the organs of the body by a system of *tracheal tubes* which begin at lateral openings (*spiracles*) and spread everywhere in the form of a branching net-work. There is thus no venous blood, as it is constantly aërated at every point. The disadvantage in this method is that such an animal must remain of small size, for by this means it is impossible to furnish fresh air in large quantities at a distance from the source of the supply. The arrangement of spiracles and tracheal tubes in the grasshopper is as follows :

I. The Spiracles (stigmata).

These are the breathing-pores through which the respiratory system communicates with the exterior. There are ten pairs of these in two lateral rows on thorax and abdomen. There are none on the head.

First Thoracic.................On membrane between pro- and meso-thorax, concealed by the posterior border of the pronotum.

Second Thoracic................On meso-thorax, just dorsal to the opening for the articulation of the second leg.

First Abdominal ....On first abdominal tergite at anterior margin of auditory membrane.

Second—eighth Abdominal. .....On the anterior ventral angle of abdominal tergites 2–8.

Make an outline drawing of the body as seen from the side. Number the somites and add the spiracles.

II. Dissection of the respiratory system.

Use a fresh specimen, killed by benzine. (If alcohol be used for killing, do not allow the specimen to remain in the liquid.) Open dorsally and pin into pan. Cover with water, as alcohol expels the air from the tubes and renders them hard to see. Look for a system of fine tubes and sacks, which appear silvery-white owing to the air confined within them. These are the tracheal tubes, or *tracheae*, which extend from the spiracles to every part. They vary greatly in size and shape, and may be thus placed under three categories :

1. *Simple tubules.* These are the most general, and in this form they ramify every organ and form a capillary net-work of extreme fineness. Different branches often anastomose with each other.

2. *Dilated tracheae.* These are local enlargements of the above and are found frequently along the main tubes. An important branch often begins as a dilated trachea.

3. *Air-sacs or vesicles.* These are sac-shaped expansions, connected with the main system by small tubes. Their tracheal nature is often emphasized by the appearance of branching tubules which spring from the air-sacs themselves.

The respiratory system is too complex to be followed throughout by dissection. The following series of dissections is given, from which the student may select for special study according to the condition of the specimen used. Make drawings.

1. *System of abdominal air-sacs.* There are five pairs of these connected with abdominal spiracles 2-6. They differ considerably in the two sexes, as follows :

Male. The last three pairs lie along the sides of the testis. Dorsally they often give rise to branches which supply the superficial portion of the testis. Ventrally they may be traced to the spiracles from which they arise.

Female. Here the sacs are very superficial, flattened, and often pinkish in color. They lie above the mass of ovarial tubules and connect with the opposite sacs by a series of fine tubes forming a diamond pattern.

2. *Thoracic air-sacs.* There are two pairs of these, of which the anterior are enormous. They lie directly beneath the pronotum, and are often seen projecting beyond its anterior edge in specimens from which the head has been violently torn. The posterior pair lie in the meta-thorax and are much smaller.

3. *The intestinal rosette.* This is a set of dilated tracheae belonging to the last three spiracles. It rests upon the dorsal side of the intestine and may be seen by parting the ovaries and removing the overlying fat. It sends branches to supply the rectum and the muscles of the rhabdites.

4. *Dorsal intestinal vessels.* These are two of the four main tubes which run longitudinally through the body. They lie upon the alimentary canal and extend from the intestinal rosette to the gastric diverticula. They are seen by parting the ovarial or spermatic tubules and reflecting these masses towards the sides. The intestinal vessels are often dilated in places and send up branches to the ovaries or testes, which lie upon them. This supply in the case of individuals with ripe ovaries, is exceedingly complicated and beautiful.

5. *Supply to the gastric diverticula.* The arrangement here is very regular and is as follows : The six intervals between the six diverticula are so arranged that two are dorsal (one median dorsal diverticulum lying between them), two are lateral and two are ventral. The dorsal intestinal vessels send out lateral branches, which, making six with the two ventral intestinal tubules, occupy the six intervals : the main dorsal intestinal vessels occupying the two dorsal intervals, the two ventral intestinal vessels, the two ventral : and the two lateral branches from the dorsal intestinal vessels the two lateral. Each of these six tubules divides anteriorly, the two branches supplying the adjacent sides of the two anterior tubules bordering the intervals. Posteriorly branches from the six tubules supply adjacent sides of the posterior diverticula.

6. *Supply to the crop.* This is the most distinct because here the silvery-white tubes lie upon a black back-ground. Four branches, two dorsal and two ventral, extend from the neck-region posteriorly and meet four which extend anteriorly. The opposing branches anastomose and supply the walls of the crop with a fine capillary net-work.

7. *The spiracular tracheae.* These are the tubules which come directly from the spiracles. From each spiracle come two, dorsal and ventral, which empty into their respective longitudinal intestinal vessel.

8. *Thorax and head.* The thorax is richly supplied with tracheal tubes, but these lie interwoven among the muscle fibres and are difficult of demonstration. From the thoracic system two large tubes, the *cephalic*, pass anteriorly through the neck into the head. The terminal branches of these are provided with very numerous small air-sacs. Of these, fifty-three were counted in one specimen.

97. *Nervous system.* This consists of two portions which are more conveniently studied in separate dissections.

1. *The ventral nerve cord.* This lies in the centre of the ventral side, and is found in the floor of the trough made by the body walls, after removal of the alimentary canal and reproductive organs. Prepare a fresh specimen by removal of these parts, or use one from which they have already been taken. Wash carefully in 30% and place in 95%. Strong alcohol whitens and hardens nervous matter. The nerve cord is easiest found in the abdomen, where it is only slightly protected by internal prolongations of the skeleton. In the thorax the cord is more carefully protected and is best dissected by tracing it up from the abdomen and removing the overlying parts. The cord will be found to consist of rounded bunches, the *ganglia* or nerve centers, connected by a median nerve cord which is in reality double, as may be proven by removing its sheath with a needle. From the ganglia fine nerves pass out to the surrounding parts. How many abdominal ganglia are there? How many thoracic? Typically there is a ganglion to each segment, but in higher forms there is more or less consolidation, especially at the end of the abdomen where there is a modification of the original segments. Draw.

2. *The " brain " (supra-oesophageal ganglion) and the other structures of the head.* This is more delicate than any of the previous preparations, but is done in the same manner. Remove the head from a large specimen, and, holding it between the thumb and finger, shave off the surface of the face with a sharp scalpel. Place it in the dissecting pan, cover with 95% and examine with the lens. Fine pins may be put through the extreme edge, to hold it in place. A whitish mass will be imperfectly seen through the irregular hole made by the scalpel. This is the *brain* or *supra-oesophageal ganglion*, and may be brought into full view by cutting away the edges of the hole and removing a mass of loose fat which often covers it. Notice the huge *optic lobes* which go to the eyes, a pair of solid cups connected by stalks : between these a pair of *ocellar lobes* with nerves to the lateral ocelli : beneath, there are a pair of *antennal lobes*, supplying nerves which run through the antennae. In the center of the mass may be seen a short bit of nerve cut off from its connection by the removal of the facial wall. This is the *nerve to the median ocellus.* Draw the brain. The brain lies dorsal to the alimentary canal and connects with the remainder of the nervous system by a pair of large nerves or *commissures* which are found at the ventral lateral corners of the brain, behind the antennal lobes. They may be followed around the oesophagus, where they unite with the *infra-oesophageal ganglion*, the first of the

ventral chain, supplying the month-parts. From this a nerve cord passes through the neck and unites with the first thoracic ganglion.

98. *Study of the minute structure of the parts (Histology).* This is done mainly by means of very thin sections artificially stained and cut by the *microtome.* The sections are to be prepared by the instructor, mainly in accordance with rules given in this book, to be learned later by the students, and distributed during the laboratory periods. The students are to bring to the microtome a clean microscopic slide (object-glass), the center of which has been smeared with a thin layer of Schällibaum's collodion fixative. The section, together with the paraffine, is placed upon this, heated gently by passing it through the top of a Bunsen gas-flame two or three times (a hot radiator is often sufficient) until the paraffine is melted. The slide is then immersed in a jar of turpentine (a few drops of turpentine may be poured over the slide instead) until the paraffine and excess of fixative are entirely removed. The slide is then to be carefully wiped with a clean cloth, leaving a small area about the section, a drop of Canada Balsam placed upon the section and a cover-glass gently laid on. Imprisoned air-bubbles will work out of the preparation in a short time. If correctly done, each portion of the section should be in the same position as when in the paraffine. If a portion floats off in the turpentine, the slide was not heated enough, or perhaps not enough fixative was used. Too thick Balsam will often push the section apart. Sections and other preparations may be made *ad libitum* of any portion but the following preparations are recommended. They are best used to accompany the dissection, but as fresh specimens are necessary for dissection, while carefully preserved material is sufficient for sectioning, it is often expedient to finish the dissection during the season in which living grasshoppers are readily obtainable.

1. *Digestive system.*

   (a) Cross-section of crop in region of teeth.

   (b) A piece of the wall mounted flat, to compare with the last. This may be done by the student, as follows: Place in three watch-crystals, 95%, 100%, and oil of cloves (or turpentine). Select the piece desired and place it 5–10 minutes in each of the fluids and in the order given. In this way the water is entirely extracted and the specimen prepared for reception into the Balsam. For this, place a drop of Balsam in the middle of a slide, and spread out the specimen in it, inner side uppermost. It is then to be covered in the usual way.

(c) Cross-section of gastric diverticula, anterior portion ; either separately or in connection with the proventicular region of the alimentary canal.

(d) Cross-section of stomach, anterior portion.

(e) Cross-section of stomach, posterior portion, including a snarl of urinary tubules, which will be cut in all planes, longitudinal, cross, etc.

(f) Cross-section of intestine.

(g) Cross-section of rectum and rectal glands.

2. *Female reproductive system.*

(a) An ovarial tubule stained in Borax Carmine and mounted whole. (Stained specimens may be given the student, which may be handled as in 1 (b).)

(b) Cross-section of a mass of tubules, which will cut them at different planes.

(c) Cross-sections of oviduct, in different places.

3. *Male reproductive system.*

(a) A spermatic tubule stained and mounted whole.

(b) A longitudinal section of a tubule showing the gradual transformation of the large cells at the apex (*spermatoblasts*) into the *spermatozoa*. Such a section is very difficult and often impossible to furnish to an entire class. A cross-section of a mass of tubules will give many details which may be interpreted by reference to a single longitudinal section.

(c) Cross-section of vas deferens.

4. *Respiratory system.*

(a) Temporary mount, unstained, of a mass of tracheal tubes, to show the spiral thread.

(b) Permanent mount of a stained tubule to show the nuclei of the intima.

5. *Nervous system.*

Sections through the brain are beautiful, and show well the arrangement of the central cells, but the brains must be dissected out and treated separately, and a single brain will not yield more than ten good sections.

6. *Muscular fibres.*

A few fibres may be isolated by picking them to pieces on a slide in a drop of water. They may then be temporarily mounted to show the striations. Stained muscular fibres will bring out the nuclei of the investing membrane (*Sarcolemma*).

Type XIX.—Mya Edulis (or Venus Mercatoria).

99. *Obtaining and preparing.* Clams may be obtained inland at the fish-markets and restaurants except during the coldest weather. The forms named above are distinguished by fishermen as "soft" and "hard-shelled clams" respectively. The first, *Mya*, is the one described here, but *Venus* will be found to correspond very well. If the clams are laid in warm fresh water for some time (the water is best left in the sun) they will become passive or dead with the parts expanded. For histological details, as with other forms of large size, the separate organs should be removed from a fresh specimen and preserved with special reference to the object in view.

100. *General dissection.* Is the animal *bilateral?* Can you find the bilateral plane? Are the shells dorsal and ventral, or right and left? Is the animal compressed or depressed? Each shell has a knob at the top, the *umbo*, from which proceed a series of concentric curves, the *lines of growth*, each representing the edge of the shell at some former time. Were the shells always of the same shape? The anterior end of the shells is more rounded, the posterior more prolonged. Back of the umbo is an elastic ligament, the *hinge ligament*, which tends to open the shells. Against this work two internal muscles, the *adductors*, which stretch across from shell to shell. When these are relaxed or cut, the shells open through the force exerted by the hinge ligament. Each shell is lined by a delicate membrane, the *pallium* or *mantle*, the edges of which project a little beyond the edges of the shell, and are grown together, except at the anterior end, where a little space is left through which projects a small muscular locomotive organ, the *foot*. Posteriorly the mantle forms a double tube, the *siphon*, which is capable of extending twice the length of the shell, or may be wholly retracted. Look for two openings at the end. The ventral one, towards the free edges of the shell, is the incurrent opening, *branchial siphon*, and the dorsal one, towards the hinge, the excurrent, *cloacal siphon*. Place the animal upon its left side (right side uppermost) and slip a scalpel along the edge close to the right shell, separating it from mantle and siphon. Then reaching in further, and

keeping close to the under surface of the shell, cut the two adductors, one
near the anterior, and one near the posterior end, lying towards the dorsal
side. Then remove the right shell entire, cutting through the hinge ligament.
Clean and dry the shell removed, and when dry notice :—The scars for the
anterior and *posterior adductor muscles;* near these, smaller scars for foot-
muscles, *protractors* and *retractors;* a line parallel to the edge of the shell
where the mantle was attached, the *pallial line,* which curves in posteriorly
to form the *pallial sinus* for the reception of the siphon : dorsally, the hinge,
where the two shells are articulated. Lift up the exposed mantle and notice
the relations of siphons, adductors, and foot. Find the opening for the foot.
The large ventral chamber now exposed is the *infra-branchial chamber* communicating with the exterior by the brachial siphon. Prove this. A large
rounded portion of the body with the foot attached, hangs free in this cavity,
upon each side of which are two delicate gill plates. In front of these are
four flat *oral tentacles* surrounding the mouth. Remove the remains of the
mantle and other overlying parts from just beneath the hinge and find the
heart, perforated by the intestine. Notice the *pericardial chamber* in which
it lies. Trace out the intestine posteriorly to the *anal opening* around the
posterior adductor muscle. With which siphonal opening does it communicate? Cut through the membrane at the base of the gills, and thus expose
the *supra-branchial* or *cloacal chamber.* Notice the perforations in its floor
for the water which comes from the gills.

101. *Dissection of special systems.* The above section serves mainly to
give the general topography of the parts, and the location of the cavities.
As both shell and mantle are formed from an integumental fold, the infra-
branchial chamber is outside the true body and hence not comparable to a
coelom. The same may be said of the supra-branchial chamber. The pendulous body mass is solid and consists of digestive tract, hepatic, and reproductive glands, packed firmly together and enclosed in a layer of muscle fibres.
The true coelom is thus limited to the pericardial chamber, which contains the
inner openings of the nephridia. After this general view, the anatomy of
each system should be studied, making use of different aspects and exposures,
using fresh specimens when necessary. For some systems, a complete ventral or dorsal view is useful, obtained by removing both shells and pinning
the animal out as desired.

1. *Respiratory system.* Each *gill-plate* is composed of two layers, connected by frequent commissures, bearing between them water-pores. The
blood circulates through blood spaces in the commissures and the lamellae

themselves. The water enters the infra-branchial chamber through the branchial siphon, passes through the water-pores into the supra-branchial chamber and out through the cloacal siphon. The incurrent water brings with it nutritive material which is taken up by the mouth, and the water in the supra-branchial chamber receives waste material from nephridia and rectum, as well as the generative products.

2. *Circulatory system.* *Heart* with median *ventricle*, perforated by intestine, and two *lateral auricles* which lie at the base of the gills and receive from them the aerated blood. From the ventricle proceed *anterior* and *posterior aortae*, which branch into arteries that supply the body. From the open termini of these, the blood passes into a *lacunar system*, through which it is conveyed to the gills.

3. *Excretory system.* The paired *organ of Bojanus* lies below and somewhat posterior to the heart. Remove the gills at their base, and open the supra-branchial chamber laterally. The organs appear as greyish-yellow masses, and show a minute opening near their anterior end, the opening of the *ureter*. The *nephridial* openings lie in the floor of the pericardial chamber (=coelom), and may be demonstrated by opening the chamber dorsally and removing the heart.

4. *Digestive system.* To dissect this, commence at the rectum and trace the canal in an anterior direction to the mouth. It consists of a long, coiled tube, running through the pendulous body, and closely packed in on all sides by the liver above and the reproductive gland below. The muscular wall of the body should be removed and the canal carefully separated from the surrounding mass. A transparent rod is generally present, lying in the intestine in the hepatic region. Its use is unknown.

5. *Reproductive system.* Bisexual, the male and female glands closely resembling each other. The *germ-glands* are very numerous whitish masses, in structure like bunches of grapes. The *terminal duct* collects the germ-cells from all the bunches, and opens into the supra-branchial chamber, at the end of a little *papilla*, which projects from the wall of the body just anterior to the opening of the ureter.

6. *Kebers Organ.* A brownish mass lying upon the mantle just under the hinge, and between the liver masses and the organ of Bojanus. Its use is unknown, but it is probably connected with circulation or excretion.

7. *Nervous system* consists of three main double ganglia connected with commissures, and of nerves which proceed from the ganglia and supply the

different parts. The *cerebral ganglia* lie above the mouth, upon the anterior
adductor muscle. The *visceral ganglia* lie upon the posterior adductor, underneath. The *pedal ganglia* lie in the anterior part of the foot. The commissures are the *cerebro-visceral*, lying superficially above the base of the gills ;
and the *cerebro-pedal*, through the visceral mass.

Type XX.—Helix Pomatia.

(This is the large vineyard snail of Europe. Any large native *Helix* may be
used instead.)

102. *Obtaining and preserving. Helix Pomatia* is used in Europe as a
common article of diet. It hibernates under leaves and stones, first closing
the mouth of its shell with a layer of lime (*epiphragma*). In this state it is
gathered in quantity and sent to the cities, and may be occasionally obtained
in the large restaurants of American cities (Philadelphia). When obtained,
break out the epiphragma and the snail will resume its activity. The common New England form, *H. albolabris*, is very typical in structure, but much
smaller and seldom found in quantity sufficient for class use. It occurs in
damp woods, crawling over moss or resting beneath stones or pieces of decayed wood. The ordinary methods of killing cannot be applied to snails, as
they will withdraw into their shells at the least alarm. To obviate this, use
the following method :—Place the snails in a jar which can be hermetically
closed, and fill it entirely with water, by immersing the whole in a pail, and
closing the jar while immersed. In this the snails will become slowly asphyx-
iated (24–36 hours) and will remain expanded, after which they can be fixed
and preserved in the usual way. Water from which the air has been expelled
by boiling will accomplish the end more speedily. The result may be also
hastened by the addition of a little chloral hydrate (2–3%).

103. *External characteristics.* An expanded animal may be seen to consist of three portions : head, foot and visceral sac. The latter is contained in
a spiral shell, which may also contain the other two when contracted. In
what respects does the shell differ from that of a clam? Is it bilateral? If
it were plastic and capable of being unrolled, could it be placed so as to become so? What would then be the shape of the visceral sac? Are the head
and foot bilateral? Is it probable that the shell and visceral sac were originally bilateral also? Is the shell carried exactly on the top of the animal, or
inclined to one side? Is this always the same side? Which? Viewing the
shell from above, it is seen to consist of a gradually expanding spiral, wound

about a center, the *apex*. Is the spiral wound towards the right (with the hands of a watch) or in the opposite direction? Are all specimens alike in this? The separate spirals are divided from each other by a crease, the *suture*. The open end of the last spiral is termed the "*Mouth*," and its free edge, the "*lip*." Notice the slightly curved lines upon the spirals, at right angles to the suture. What is their significance? Do they correspond to anything on the surface of a clam shell? On the head notice :—The *long tentacles* (*ommatophores*) which bear at their tips the eyes and probably the olfactory organs ; the *short tentacles* ; the *genital opening* on the right side, at the base of the ommatophore (behind the short tentacle in *H. albolabris*) ; the *mouth*, surrounded by soft lobes, the *lips*. Probe the mouth gently with a needle and feel a hard chitinous dorsal piece, the *jaw* or *maxilla*. Beneath the margin of the shell on the right side, notice an opening, the *breathing pore* or *pneumostome*, which opens into a large respiratory sac, the so-called "lung." The intestine runs along the wall of this sac and opens at the edge of the pneumostome. The nephridial opening is at the same place, but cannot be demonstrated from the exterior. During life the pneumostome is under the control of muscles. The foot is provided with a flat creeping sole, posteriorly prolonged, and running anteriorly beneath the head, nearly as far as the mouth.

104. *Dissection.* Carefully remove the shell, beginning at the edge and taking it away in little pieces. Notice that the shell has a central support, the *columella*, about which the visceral sac is wound. Commencing below, uncoil the visceral sac, removing every vestige of columella and shell. The entire visceral sac is covered by a delicate membrane, the *mantle*, which encloses the organs, and has a thickened edge below, the *mantle ridge*, which marks the boundary of the shell. Pass a probe into the pneumostome. It enters the respiratory chamber, and may be seen through its transparent wall. Open this chamber with the scissors, and notice upon the right side :—the *heart*, enclosed in a delicate sac, the *pericardium* ; the *nephridial organ* with its duct, and the *rectum*, terminating in the *anus*. A rich net-work of blood-vessels is spread over the walls. The floor of the lung-chamber is formed by the body-wall, which covers the visceral organs and thus shuts off the respiratory chamber from the coelom. (cf. the branchial chambers of the cray fish and the clam.) For farther dissection make a median dorsal incision, beginning just above the mouth. Carry it posteriorly through the mantle-ridge and the floor of the respiratory chamber. As it is difficult to continue a mid-dorsal incision upon the coiled visceral sac, the entire mantle may be removed from this, thus liberating the organs. All the cavities of the body are thus opened and the organs exposed for study. The specimen should be pinned down in a dissecting pan, the cut edges of the body wall drawn apart, and the separate systems studied. The parts are to be identified by the synopsis given below, the details of form and relationship are to be left to the student. It is well to notice the following points : The *alimentary canal* may be isolated first, following the windings of the intestine to the apex of the spirals and back to the respiratory chamber. The large brown organ filling the spirals is the *liver*, and should be removed with the canal. Take great care and not harm the *hermaphroditic gland*, or its *duct*, lying in close connection with the liver and intestine near the apex. Observe the *central ganglia* of the nervous system lying upon the *pharynx*, and do not investigate the interior of the pharynx until the nervous system has been studied. When this is done, remove the pharynx for detailed study. Find the *maxilla* and the *radula*, a tough membrane covered with countless minute teeth, in the interior of the pharyngeal cavity. This should be removed and mounted in Balsam.

105. *Anatomical Synopsis.*

1. *Digestive system. Pharynx*, an oval organ with very muscular walls, in which lie the *maxilla* and the *radula*—Short *oesophagus* expanding into the large fusiform *stomach—Salivary glands*, lying on each side of the stomach, with long *ducts*, which open into the pharynx—*Intestine*, which shows an enlarged area where the *ducts of the liver* enter—*Liver*, several large dark-brown lobes, closely packed in the upper part of the spiral—*Anus*, in the wall of the respiratory chamber, close to the pneumostome.

2. *Reproductive system. Hermaphroditic gland* and *duct* at apex of sac *albuminous gland*, a large white mass with a free end—*uterus* or *oviduct*, a somewhat coiled crenate tube, accompanied on one side by a smaller tube, the *vas deferens*—This latter leaves the oviduct, makes a turn upon itself and expands into a sheath which contains the *penis* or organ of copulation. The sheath opens at the genital orifice—The *flagellum*, a gland in the form of a long free tube at the inner end of the penis-sheath—At end of oviduct a sac and two bunches of finger-like glands. The sac contains a spicule which aids in copulation, the so-called " Cupid's arrow " (Liebespfeil) ; the glands probably assist in the formation of the egg-shell—The *spermatheca*, a spherical sac, communicating by a long slender duct with the lower portion of the oviduct.

3. *Circulatory and respiratory systems.* *Heart,* with one *atrium* and one *ventricle,* distinguished by the thickness of their walls. The circulation is partly in blood vessels, and partly lacunar, there being no capillaries. The blood from the coelom and other blood sinuses is brought into the respiratory chamber by a *ring-vessel,* which runs along the mantle-ridge and branches into a net-work of fine vessels. These are re-collected into a *pulmonary vein,* which conveys the now aerated blood to the atrium. A single aorta passes from the ventricle, but soon gives off the *visceral artery* towards the apex of the visceral sac. The aorta then passes into the body cavity. Its main branches are *intestinal, genital, salivary* and *pedal,* to the corresponding parts.

4. *Nervous system.* The three typical ganglia, as seen in the clam, are here united into a complex lying upon the pharynx. The *cerebral* ganglion is anterior, connected by a double commissure to the *viscero-pedal* ganglion just posterior to it. This latter is perforated by an artery which divides it into an upper (=*pedal*) and a lower (=*visceral*). Anterior to the cerebral is a small pair, attached by commissures, the *stomato-pedal.* The following are the main nerves:— From cerebral ganglion : *tentacular* (to ommatophores) *outer-labial, inner-labial, facial* (to front of head and short tentacles) *auditory* (to ear-sacs on side of pedal ganglion) a median nerve to the penis. From viscero-pedal : *genital* (to reproductive organs and heart), *anterior, middle* and *posterior mantle nerves.* A large number to foot, mostly small.

Type XXI.—Bugula sp?

106. *Occurrence. Preservation. Bugula* is a colonial polyp closely resembling a hydroid, and found on all sorts of submerged and floating objects, especially on eel-grass. Under the lens a branch resolves itself into a colony of sessile animals living in cells, into which they are capable of entirely withdrawing. As the skeletal portion is very durable, a great many of the branches collected will be found partially or wholly empty, and thus each specimen must be examined before preserving. They should be treated as in the case of hydroids in order to prevent the retraction of parts.

107. *Study of a stained specimen.* Select a good branch fixed with parts expanded, and stain and mount entire. Notice the separate cups or cells, *zooecia,* in which the separate animals live. How are they arranged, rela-

10

tively? Look along the edges at the bases of the zooecia, for little pairs of jaws, something like birds' beaks, mounted on pedicles. These are the *avicularia*, probably specialized individuals, fitted to detain and hold animals which will attract the small forms upon which Bugula feeds. At the bases of the separate zooecia are sometimes found oval cysts containing a single egg. These are *ooecia*, which may be another specialized form, or have the worth merely of a brood capsule. The animal itself bears a series of tentacles, upon a semi-circular ridge, the *lophophore*. How many tentacles are there? They differ from those of hydroids in the presence of cilia and in the absence of cnidoblasts. The alimentary canal consists of *pharynx*, *stomach* with a terminal *coecum*, *intestine* and *rectum*. As in most sessile forms the canal curves back upon itself, the *anus* lying just outside the lophophore. A muscular cord, the *funiculus*, extends from the coecum to the bottom of the cup, probably to prevent over-extension of the parts. Ovaries develop on the body-wall near the coecum. Testes appear upon the funiculus. The polyps are hermaphroditic.

108. For special examination of the skeleton, soak a colony in K-O-H, wash in water and examine. This will bring out the details of the avicularia, etc.

Type XXII.—Asterias vulgaris.

109. *Collecting and preserving.* This is the common star fish, abundant everywhere at the sea-coast and obtained at low tide, especially about mussel-beds and the piles of wharves. They are preserved for ordinary dissection by being put alive into 95%, as they are porous enough to allow free entrance of the fluid, and contain sufficient water to dilute the alcohol. For histology the separate parts should be dissected out and preserved.

110. *External characteristics.* Is the star-fish bilaterally symmetrical? How many planes will divide the animal symmetrically? Such a structure is termed radiate. Upon what number is the radial symmetry based? What is the usual number among Cnidaria? The body is divided into a central portion or *disc*, and five triangular projections, the *rays*. The distinction between the two is slight here, but very definite in the "serpent-stars" (Ophiuridea). Compare the two surfaces; the flatter of the two is beneath when the animal crawls, and has the *mouth* in its centre, and thus is called the *oral* surface. The other is the *ab-oral* surface. *Oral side:*—Notice the central mouth, its shape—very numerous soft tentacles, the *ambulacral feet*, extending through the center of each ray as far as the tip. How many rows are there? Cut them entirely away from a small area and notice the pores in the floor of the *ambulacral groove*, through which they are projected. These are the *ambulacral pores*. Have they a definite arrangement? In another area, part the rows of feet in the middle and expose a central ridge, the *nervous cord*, which proceeds from a central ring. Follow this ont to the point of a ray and note the *pigment eye*. *Ab-oral side:*—Notice the plates beset with spines. Are there soft spaces between these plates? Notice often in these, projections called *papulae*, very thin portions of the integument which project out between the spines, and in the living animal are very sensitive and capable of sudden withdrawal. These, as well as the ambulacral feet, are used in respiration, as a special respiratory system fails. Search with a lens among the spines for minute organs, appearing like little knobs mounted on exceedingly delicate stalks. These are the *pedicellariae*, and appear under the microscope as tiny forceps, some two- and some three-tined. They are used to pick grains of sand and all forms of debris from among the spines. Treatment with K O-H will resolve the skeletal structure of these. Upon the disc, at one side notice a raised circular plate, the only departure from the radial symmetry. It is the *madrepore plate*, a porous structure through which the water passes in to the ambulacral system (see below).

111. *Dissection.* The entire ab-oral surface should be removed, cutting through the edges of the rays with the scissors, and entirely around the madrepore plate, leaving it with the preparation. The disc is filled with the *stomach*, a delicate membrane, much folded and wrinkled. Five lobes may be found, which pass out a short way into the rays ; above, a very small and tapering intestine leads to a minute anus, cut off by the removal of the disc. In each ray are a pair of brownish tree-like structures, the *hepatic organs*, each of which possesses a central duct. The two ducts in each ray unite into a median duct, which opens into the gastric diverticulum of that ray. At the base of each ray are two sexual glands (very small in immature specimens) the ducts from which unite in such a manner that the glands of two adjacent rays open by a common orifice in the interradial angle. The most characteristic system is the *water vascular* or *ambulacral*. The madrepore plate is continuous with a sinuous canal, the *stone canal*, which may be followed into a circular ridge. This ridge conceals the *ambulacral ring*, from which five *radial ambulacral vessels* proceed through the center of each ray The course of this canal is also through skeletal pieces, and is marked by a median ridge in the floor of each ray. From this canal, very numerous lateral branches supply little oval vesicles, the *ampullae*, of which the ambulacral feet are the continuation. The entire apparatus forms a system by means of which water may be forced into and out of the ambulacral feet, causing them to be pushed out and withdrawn. The relation of these parts is best understood through the aid of cross-sections of a ray [113].

112. *Study of skeletal plates.* Cut off a ray and boil it in a test tube of K-O-H until it falls apart. The little plates composing the skeleton are separated by this means. Wash several times by decantation and finally bring the whole into a Stender dish for study. Compare these with another specimen, not boiled as much, and in which the plates still hold together. The procedure can be regulated as desired by different degrees of dilution of the reagent, and by varying the time of boiling. Obtain by these means an accurate idea of the architecture of the skeletal plates, and express the result by a diagram.

113. *Cross-section of a ray.* Select for this a few very small star-fish and preserve by warm alcoholic sublimate. When sections are desired, remove a ray and decalcify by letting it stand several days (as long as necessary) in a dish of picric acid, to which has been added a trace of nitric acid. The acid should be daily changed and the process continued until the specimen is no longer gritty when tested with a needle. Wash out in 70%, and stain (2-3 days) in Alum Carmine. Dehydrate, imbed and section as usual.

114. *Larval forms.* These are captured by the tow net. They are the famous *bipennariae* and *brachiolariae*. They may be investigated as in the case of other forms found in tow (nauplias, etc.) and may be mounted either stained or unstained. Handle with a pipette, or upon a slide.

TYPE XXIII.—Arbacia punctulata.
Strongylocentrotus Drobachiensis.

115. *Occurrence and preservation.* These are the "sea-urchins" common on every coast. *Arbacia* is the small dark-purple form, found from Massachusetts Bay southward, and *Strongylocentrotus* is the larger olive-green species, with smaller spines, common farther north. Either species will do, the minor differences making the comparison interesting. They may be preserved as in the case of Asterias, but an artificial opening should be made into the interior to allow the passage of the alcohol, as the skeleton is much firmer than in the star-fish. A slit through the *peristome*, along the edge of the test, is the most practical.

116. *External anatomy.* Study and explain the symmetry of the animal. What is the shape of the shell? Find the oral and ab-oral aspects. *Oral side :*—In the center, the circular *lip*, surrounding five *teeth*, set in hard jaws. The soft area continuous with the lip is the *peristome*, reaching to the margin of the shell or *test*. Upon the peristome are ten large *ambulacral tentacles*, surrounded by *pedicellariae*. At the border of the test are the oral ends of five rows of *ambulacral feet*. Trace these around upon the ab-oral side. *Ab-oral side :*—In the center, the *anus* protected by four *anal plates* (in *Strongylocentrotus* the four plates are replaced by numerous little oval plates). Surrounding the anal plates are five large *genital plates*, alternating with five small *ocellar plates*. One of the genital plates includes the *madreporic area*. Find the genital openings and the eye spots in their respective plates. What relation do these plates sustain to the rows of ambulacral feet?

Spines and test :—Notice the spines. Tear one away from the test and notice the articulation. Have the spines any regular arrangement? Can the position and size of the spines be determined on a denuded test? Beginning at the anus, pull off the spine from a considerable area, including a row of ambulacral feet and the entire space between two rows. Notice that the position of the ambulacral feet is indicated by two rows of perforated plates, while the space between two ambulacral rows is filled by two rows of larger plates. The area of the feet is termed the *ambulacral area*, and the space between them the *inter-ambulacral area*. Which bears the largest spines? What is the arrangement of the ambulacral feet (as determined by the ambulacral pores)? What is the total number of rows of plates running meridionally around the test? [After the dissection, the test may be cleaned by boiling in dilute K-O-H, and the separate plates isolated. If a perfect test be taken, the separate skeletal elements may be mounted on a black card as in the diagram of LANG].

117. *Dissection.* This should be performed from the oral side. Remove the peristome and crack away the margin of the test as far as desired. In the centre lie the teeth and jaws, forming a symmetrical cone-shaped mass, "*Aristotle's lantern.*" Lift this carefully and notice that the alimentary canal runs through it, the free part beginning at the center of its base. Separate and study the parts of the "lantern." It is moved by muscles attached to the edge of the test, and from one part to another of the lantern itself. The internal anatomy is similar to that of *Asterias*, presenting, however, interesting minor differences in every system. Dissect it carefully and compare step by step.

118. *Study of larvae.* These are found in tow and are to be treated as in 114. The larva is termed the *Pluteus*, similar to that of Ophiuridea, but distinguished from that of other Echinoderms by the great length of the lateral arms.

Type XXIV.—Balanoglossus sp?

119. *Occurrence.* This is one of the numerous forms of mud-worms or clam-worms, found in mud-flats at low tide. It formerly received but little attention, but has lately become an important form, on account of its probable kinship to Tunicata, and hence its relationship to Vertebrates. According to some it possesses also affinity to the Echinoderms. Some species develop directly, others by a free-swimming pelagic larva, the *Tornaria*, similar to the larvae of Echinoderms.

120. *External anatomy.* The body is bilateral and divided into three regions, *proboscis, collar* and *body* proper. The first two are rounded, the body somewhat depressed. The *dorsal side* is marked by the presence of *paired gill-slits*, which lead from the alimentary canal, and open dorsally by a series of *pores*. A median stripe upon both dorsal and ventral surfaces marks the position of *median blood-vessels* and *median nerve cords*. The *mouth* leads in from the anterior edge of the collar upon the ventral side. The *anus* is situated at the posterior extremity.

121. *Internal anatomy.* Owing to the small size of the animal, and the partial filling up of the coelom with muscular fibres and connective tissue, simple dissection is impracticable. The parts and their relations must be studied by sections, the preparation of which is rendered difficult by the peculiar consistency of the animal. Treated in the usual way, the tissues are apt to harden to the consistence of horn, and become impossible to cut. Better results may be obtained by imbedding in *celloidin*. The most useful sections are cross-sections through the lower portion of the proboscis, the collar and the branchial region of the body; and a median longitudinal section cut dorso-ventrally through the anterior portion. The following synopsis of the systems may be helpful :

1. *Digestive system.* A nearly straight canal, running through the entire body, from the ventrally situated *mouth* at anterior edge of collar, to the terminal *anus*. Of morphological importance is an anteriorly extended *dorsal diverticulum* in the region of the collar, considered by some the homologue of

the *notochord* of the Chordata. A *skeletal body*, probably secreted by this last, lies between it and the dorsal-wall of the alimentary canal, and sends two processes around the canal itself. This is important as showing a skeletogenous quality in the diverticulum.

2. *Circulatory system.* Two median vessels, dorsal and ventral, which give off lateral twigs. At the anterior end of the dorsal vessel, lying upon the diverticulum, the vessel is enlarged, forming what KOHLER and BATESON call the "heart."

3. *Respiratory system.* The "gills" consist of paired diverticula of the dorsal-wall of the intestine opening *externally* through pores in the dorsal integument. The blood passes to these either through the lateral branches above given, or from a pair of lateral vessels (KOWALEVSKY).

4. *Reproductive system.* Bisexual ; the two sorts of glands closely resembling each other. They consist of *paired glandular pockets* opening separately by pores on the dorsal side. Some of these lie in the gill-region, but are generally best developed in the region just posterior to this.

5. *Excretory system.* This function has been ascribed by BATESON to the *dorsal pore*, a minute tube passing into the proboscis from a median dorsal opening just beneath the edge of the collar. According to others (SPENGEL), the dorsal pore is a means by which water passes into the proboscis, making the latter an *ambulacral organ*.

6. *Nervous system.* The central organ is a thick cord lying dorsally in the collar region, from which extends a mid-dorsal cord through the body, accompanying the dorsal blood-vessel. Immediately behind the collar the dorsal cord gives off two lateral commissures, which unite ventrally in the region of the first pair of gill-diverticula, and form a mid-ventral cord, which follows the blood-vessel of the same name.

122. *Study of Tornaria larva.* Certain species of Balanoglossus possess a free-swimming larva, the *Tornaria*, occurring in oceanic tow collected near the regions inhabited by the adult. They are studied and prepared in the same manner as other minute larvae.